I0467871

Methods to Characterize Environmental Settings of Stream and Groundwater Sampling Sites for National Water-Quality Assessment

By Naomi Nakagaki, Kerie J. Hitt, Curtis V. Price, and James A. Falcone

National Water-Quality Assessment Program

Scientific Investigations Report 2012–5194

U.S. Department of the Interior
U.S. Geological Survey

U.S. Department of the Interior
KEN SALAZAR, Secretary

U.S. Geological Survey
Marcia K. McNutt, Director

U.S. Geological Survey, Reston, Virginia: 2013

For more information on the USGS—the Federal source for science about the Earth, its natural and living resources, natural hazards, and the environment, visit http://www.usgs.gov or call 1–888–ASK–USGS.

For an overview of USGS information products, including maps, imagery, and publications, visit http://www.usgs.gov/pubprod

To order this and other USGS information products, visit http://store.usgs.gov

Foreword

The U.S. Geological Survey (USGS) is committed to providing the Nation with reliable scientific information that helps to enhance and protect the overall quality of life and that facilitates effective management of water, biological, energy, and mineral resources (http://www.usgs.gov). Information on the Nation's water resources is critical to ensuring long-term availability of water that is safe for drinking and recreation and is suitable for industry, irrigation, and fish and wildlife. Population growth and increasing demands for water make the availability of that water, measured in terms of quantity and quality, even more essential to the long-term sustainability of our communities and ecosystems.

The USGS implemented the National Water-Quality Assessment (NAWQA) Program in 1991 to support national, regional, State, and local information needs and decisions related to water-quality management and policy (http://water.usgs. gov/nawqa). The NAWQA Program is designed to answer" What is the quality of our Nation's streams and groundwater? How are conditions changing over time? How do natural features and human activities affect the quality of streams and groundwater, and where are those effects most pronounced? By combining information on water chemistry, physical characteristics, stream habitat, and aquatic life, the NAWQA Program aims to provide science-based insights for current and emerging water issues and priorities. From 1991 to 2001, the NAWQA Program completed interdisciplinary assessments and established a baseline understanding of water-quality conditions in 51 Nation's river basins and aquifers, referred to as Study Units (http://water.usgs.gov/nawqa/studies/study_units.html).

National and regional assessments are ongoing in the second decade (2001–2012) of the NAWQA Program as 42 of the 51 Study Units are selectively reassessed. These assessments extend the findings in the Study Units by determining water-quality status and trends at sites that have been consistently monitored for more than a decade, and filing critical gaps in characterizing the quality of surface water and groundwater. For example, increased emphasis has been placed on assessing the quality of source water and finished water associated with many of the Nation's largest community water systems. During the second decade, NAWQA is addressing five national priority topics that build an understanding of how natural features and human activities affect water quality, and establish links between sources of contaminants, the transport of those contaminants through the hydrologic system, and the potential effects of contaminants on humans and aquatic ecosystems. Included are studies of the fate of agricultural chemicals, effects of urbanization on stream ecosystems, bioaccumulation of mercury in stream ecosystems, effects of nutrient enrichment on aquatic ecosystems, and transport of contaminants to public-supply wells. In addition, national syntheses of information on pesticides, volatile organic compounds (VOCs), nutrients, trace elements, and aquatic ecology are continuing.

The USGS aims to disseminate credible, timely, and relevant science information to address practical and effective water-resource management and strategies that protect and restore water quality. We hope this NAWQA publication will provide you with insights and information to meet your needs, and will foster increased citizen awareness and involvement in the protection and restoration of our Nation's waters.

The USGS recognizes that a national assessment by a single program cannot address all water-resource issues of interest. External coordination at all levels is critical for cost-effective management, regulation, and conservation of our Nation's water resources. The NAWQA Program, therefore, depends on advice and information from other agencies— Federal, State, regional, interstate, Tribal, and local—as well as nongovernmental organizations, industry, academia, and other stakeholder groups. Your assistance and suggestions are greatly appreciated.

William H. Werkheiser
Associate Director for Water

Contents

Figures

Conversion Factors

SI to Inch/Pound

Multiply	By	To obtain
Length		
centimeter (cm)	0.3937	inch (in.)
meter (m)	3.281	foot (ft)
Area		
square kilometer (km^2)	0.3861	square mile (mi^2)
Mass		
kilogram (kg)	2.205	pound avoirdupois (lb)

Vertical coordinate information is referenced to the North American Vertical Datum of 1988 (NAVD 88).

Horizontal coordinate information is referenced to the North American Datum of 1983 (NAD 83).

Acronyms and Abbreviations

AML	Arc Macro Language
GIS	geographic information system
LULC	Land Use and Land Cover
LUS	Land-use study (of shallow ground water)
MAS	Major aquifer study
NAWQA	National Water-Quality Assessment (program)
NLCD 2001	National Land Cover Database 2001
NLCD 2006	National Land Cover Database 2006
NWIS	National Water Information System
RASA	Regional Aquifer-System Analysis (study)
STAID	station-identification (number)
SSURGO	Soil Survey Geographic (database)
STATSGO	State Soil Geographic (database)
SUCODE	unique identifier for a network of NAWQA sampling sites

Organizations

USGS	United States Geological Survey

Methods to Characterize Environmental Settings of Stream and Groundwater Sampling Sites for National Water-Quality Assessment

By Naomi Nakagaki, Kerie J. Hitt, Curtis V. Price, and James A. Falcone

Abstract

Characterization of natural and anthropogenic features that define the environmental settings of sampling sites for streams and groundwater, including drainage basins and groundwater study areas, is an essential component of water-quality and ecological investigations being conducted as part of the U.S. Geological Survey's National Water-Quality Assessment program. Quantitative characterization of environmental settings, combined with physical, chemical, and biological data collected at sampling sites, contributes to understanding the status of, and influences on, water-quality and ecological conditions. To support studies for the National Water-Quality Assessment program, a geographic information system (GIS) was used to develop a standard set of methods to consistently characterize the sites, drainage basins, and groundwater study areas across the nation. This report describes three methods used for characterization—simple overlay, area-weighted areal interpolation, and land-cover-weighted areal interpolation—and their appropriate applications to geographic analyses that have different objectives and data constraints. In addition, this document records the GIS thematic datasets that are used for the Program's national design and data analyses.

Introduction

Background

In 1991, the U.S. Geological Survey (USGS) initiated the National Water-Quality Assessment (NAWQA) Program to assess the quality of our Nation's streams and groundwater. Information analyzed for the NAWQA Program includes chemical concentrations in water, sediment, and aquatic-organism tissues (Gilliom and others, 2001; Gilliom and others, 1995). Over 14 million samples have been collected for the Program at more than 7,700 stream sites and 9,400 wells (U.S. Geological Survey, 2010). The geographic distribution of NAWQA's surface-water and groundwater sites sampled in the conterminous United States is shown in figure 1.

Water-quality findings are used with natural and anthropogenic characteristics for stream and groundwater sampling sites to evaluate how environmental settings affect water quality. For example, high concentrations of atrazine observed at a stream sampling site can correlate with the intense use of atrazine on cropland and poorly drained soils in its drainage basin. Characteristics of sampled sites are not only used to evaluate current water-quality conditions but also in statistical models for predicting concentrations of contaminants in unmonitored areas (Stone and others, 2008; Gilliom and others, 2006; Nolan and Hitt, 2006; Nowell and others, 2006; Stackelberg and others, 2006).

The study design of the NAWQA Program is uniform in its sampling and analytical methods, which makes it equally important to maintain data consistency in compiling site and study-area characteristics. Site and study-area characteristics in the conterminous United States were developed systematically to prevent bias introduced from applying multiple methods and multiple data sources that were compiled differently for a particular characteristic. A GIS was used to construct nationally consistent characteristics by using national geospatial thematic datasets and a standard set of methods.

Purpose and Scope

The objective of this report is to describe three methods used to characterize the environmental settings of NAWQA's stream and groundwater sampling sites and associated study areas. The three methods are simple overlay, area-weighted areal interpolation, and land-cover-weighted areal interpolation. Fictitious data values are used in most of the examples to simplify the calculations. The methods were incorporated into algorithms by using Esri's GIS software, originally in Arc Macro Language (AML) for use in the ArcInfo workstation environment, and more recently in Python for use as tools in the ArcGIS Desktop environment. The ArcGIS tools have been published and are documented separately (Price and others, 2010).

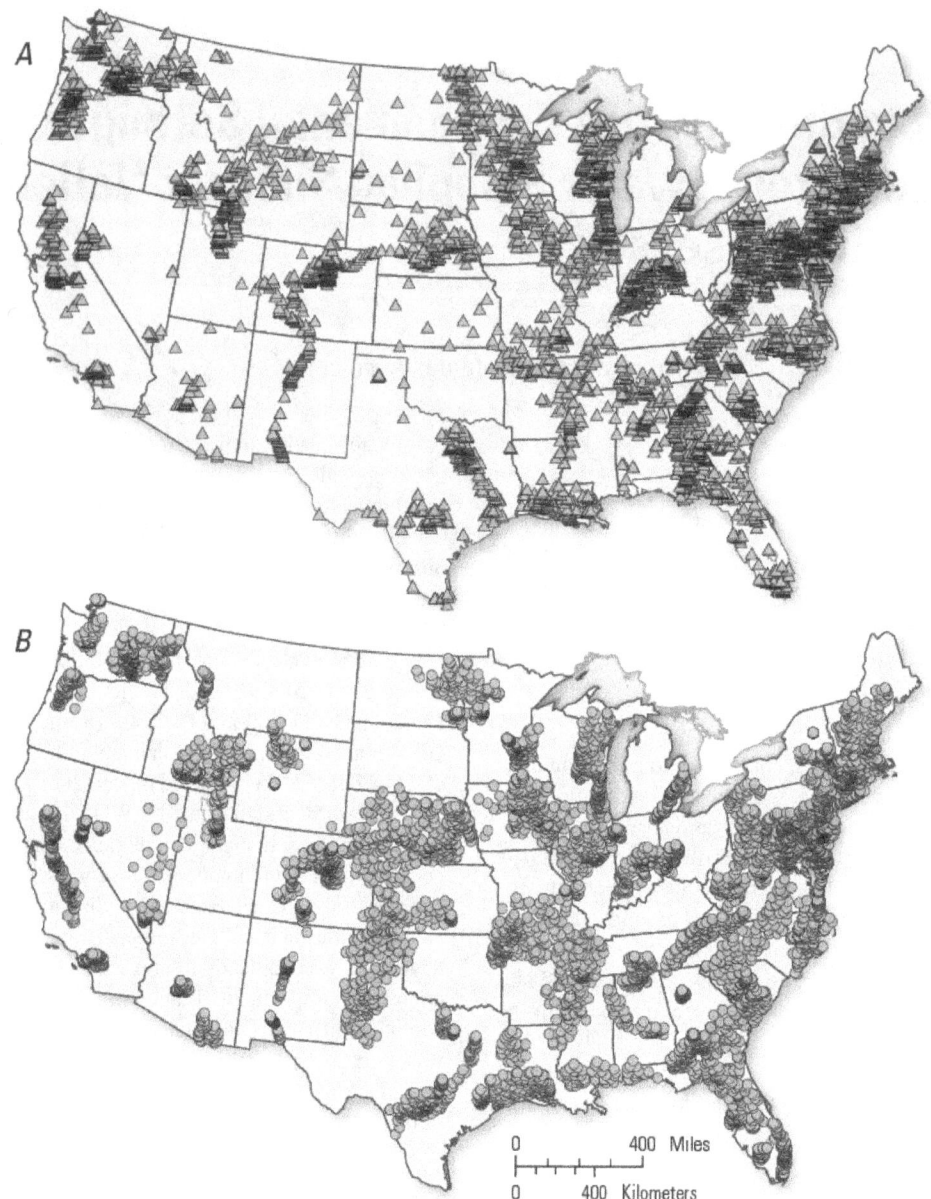

Figure 1. National Water-Quality Assessment (NAWQA) Program's (*A*) stream sampling sites and (*B*) groundwater sampling sites in the conterminous United States (U.S. Geological Survey, 2010). These maps are based on the site information data from the NAWQA Data Warehouse (Bell and Williamson, 2006), as of June 3, 2010.

The examples that illustrate the methods use GIS thematic raster data that cover the entire nation. Subsequently, the examples refer to characteristics of raster data, such as grid-cell counts for the conterminous U.S., though the methods can be used with vector data for selected regions. The data format is independent of the method, and the only requirement of the source data is the geographic coverage of the sites and study areas of interest. In the NAWQA Program, it is essential that national thematic datasets be used for characterizing sites and study areas because this helps ensure consistent and comparable characteristics across the country.

The appendix in this report provides descriptions and references for obtaining the national GIS thematic datasets used by NAWQA. It contains over 110 data sources for the conterminous United States and identifies which of the three methods discussed in this report were applied to the dataset. Most of the national GIS thematic datasets were not available for areas outside the lower 48 states; accordingly, NAWQA sampling sites and study areas in Hawaii and Alaska were not characterized. The appendix is a snapshot of the national GIS thematic datasets used by the Program and will likely be updated as newer and enhanced data sources are developed and used for geoprocessing.

GIS Datasets

This section describes the GIS datasets used to characterize NAWQA's sampling sites and associated study areas.

Sampling Sites

NAWQA sampling sites are either specific locations on streams or, in the case of groundwater, specific wells. Each NAWQA sampling site is identified by a station-identification number (STAID), which is the 8-15 digit numeric identifier for the station used within USGS's National Water Information System (NWIS) database (USGS's NWIS web interface is publicly accessible at http://waterdata.usgs.gov/nwis). The latitude and longitude of each NAWQA site, along with other site information stored in NWIS, were used to generate a national GIS point dataset. The environmental characteristics for the stream sampling sites and wells are identified by the USGS STAID.

Drainage Basins

The drainage basin for a stream sampling site is the topographically defined area of land that potentially contributes runoff to the stream. The stream sampling site and the GIS drainage basin for the USGS station, Pudding River at Aurora, Oregon (STAID 14202000) is shown in figure 2. The drainage basin for this site (hereafter referred to as the "Pudding River basin"), like most other basins, represents the contributing drainage area. Some atypical drainage basins include closed basins—basins that have no natural outlets—or areas that are considered noncontributing except during storms.

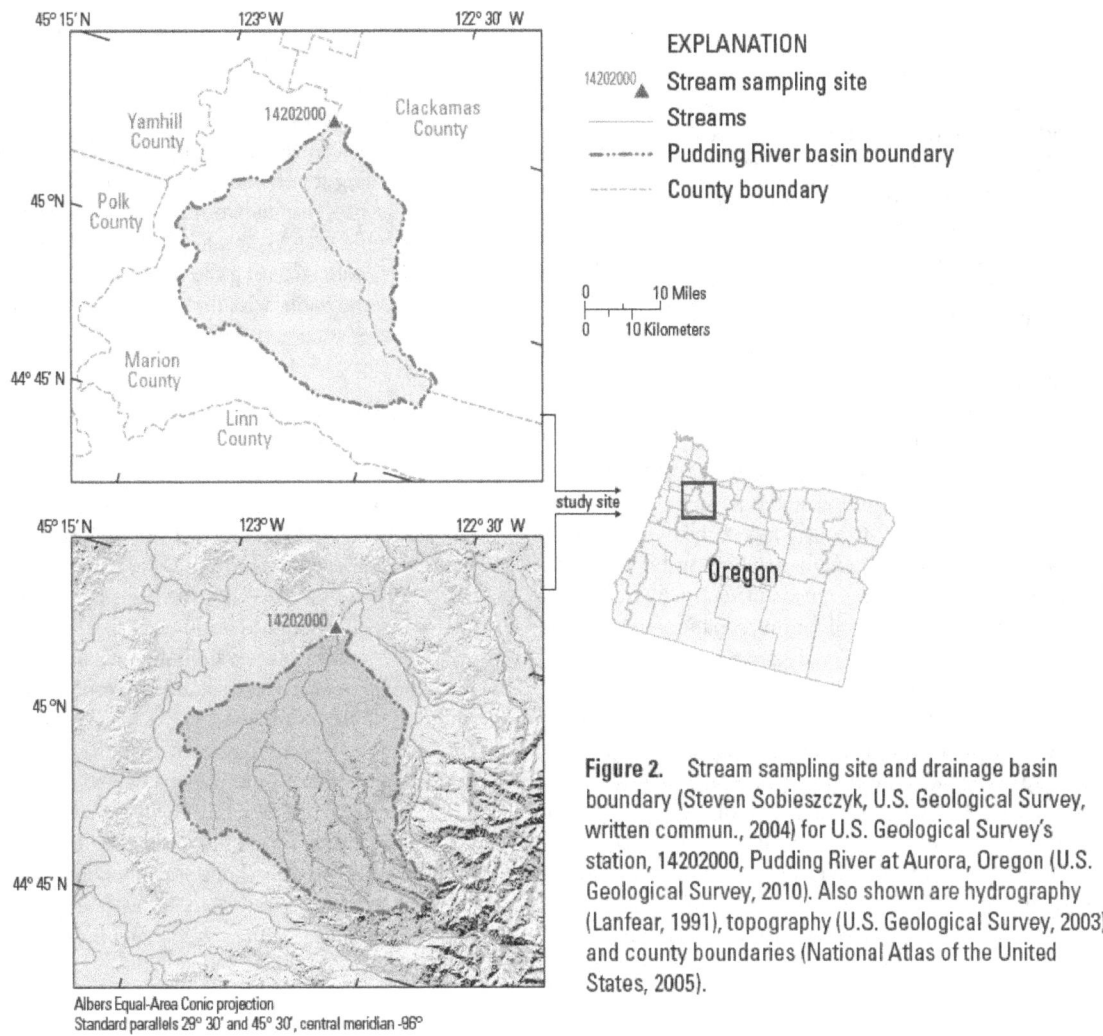

Figure 2. Stream sampling site and drainage basin boundary (Steven Sobieszczyk, U.S. Geological Survey, written commun., 2004) for U.S. Geological Survey's station, 14202000, Pudding River at Aurora, Oregon (U.S. Geological Survey, 2010). Also shown are hydrography (Lanfear, 1991), topography (U.S. Geological Survey, 2003), and county boundaries (National Atlas of the United States, 2005).

Drainage basins for the NAWQA stream sites were delineated by many USGS hydrologists and geographers from a variety of analog and digital data sources ranging in scale from 1:24,000 to 1:250,000, depending on the size of the basin (Nakagaki, 2010; Nakagaki and Wolock, 2005). The drainage basins, which were originally delineated as vector polygons, were converted to rasters at the 30-meter (m) resolution prior to geoprocessing basin characteristics. Drainage basins for the NAWQA surface-water sampling sites vary greatly in size, ranging from less than one square kilometer (km^2) to more than 220,000 km^2.

A sub-area of the drainage basin, for which a few characteristics are calculated, is defined as the "riparian buffer zone." This sub-area is defined as all areas within 50 or 100 m of any stream centerline in the basin. The characteristics of the riparian buffer zones as well as the drainage basins are identified by STAID.

Groundwater Study Areas

NAWQA's groundwater sampling sites (or wells) are characterized in two ways: by buffers around each site and aquifer boundaries linked to groups of wells (or well networks). The circular buffers, which are identified by STAID, are defined by a 500-m radius around the site and have a total area of about 0.78 km^2. No provision is made for overlapping buffers; each one is treated separately. NAWQA groundwater sampling sites processed for well networks are characterized by study areas defined by aquifer boundaries or a subarea of the aquifer underlying a targeted land use. Boundaries of aquifers and land-use study areas represent the geographic location of NAWQA's groundwater assessments, and were initially used to constrain the random selection of groundwater sampling sites. NAWQA's groundwater assessments comprise major aquifer studies (MAS) and land-use studies (LUS; Gilliom and others, 1995).

The objective of a MAS is to assess the overall quality of an aquifer that is important for drinking-water supply. The spatial extent of a MAS covers the entire hydrogeologic setting of the aquifer selected for investigation, irrespective of land use. Typically 20–30 wells are randomly selected throughout the aquifer for sampling. For example, the Biscayne aquifer study in southern Florida sampled a network of 30 wells, which provided a broad overview of groundwater quality reflecting a mixture of land uses (fig. 3A). To characterize a MAS, the overall extent of the aquifer selected for study was used as the groundwater study-area boundary (Biscayne aquifer, fig. 3A). The area of a MAS ranged from about 250 km^2 to 178,000 km^2.

The objective of a LUS is to determine the effects of a certain land use–most frequently agricultural or urban–on predominantly shallow, recently recharged groundwater. This type of study typically involves about 20-30 randomly-selected wells only in areas associated with the land use of interest. For example, a subset of the Biscayne aquifer beneath urban land in Broward County, Florida, was sampled using a network of nearly 40 wells to assess the quality of shallow groundwater within an urban landscape (fig. 3B). To characterize a LUS, the overall extent of the aquifer selected for study, and the extent of the subset of the aquifer where the targeted land use intersects the aquifer (targeted urban land in Broward County, shown in fig. 3B), were used as the groundwater study-area boundaries. The area of a LUS ranged from about 20 km^2 to 12,700 km^2, but most were less than 1,000 km^2.

Boundaries of aquifers and land-use study areas within aquifers were delineated by USGS hydrologists and geographers using various geologic, topographic, hydrologic, physiographic, and soils maps (Squillace and Price, 1996). These maps came from many sources, including the Groundwater Atlas of the United States (U.S. Geological Survey, 2000), USGS Regional Aquifer-System Analysis (RASA) studies (Sun and others, 1997), and various land use and land cover datasets. The boundaries of aquifers and land-use study areas, which are often composed of multiple discontinuous areas, were delineated originally as vector polygons then converted to rasters at 30-m resolution prior to geoprocessing for characterization. The characterization of the LUS and MAS well network is identified by an NWIS network name specific to NAWQA networks (SUCODE), which is derived by concatenating the four-letter abbreviation for the NAWQA study area with the code designating a particular groundwater study network.

Thematic Datasets

The requirements for the GIS thematic datasets for NAWQA are that the data were systematically compiled and are available for the conterminous United States. These requirements are necessary in order to develop consistent characteristics for NAWQA sites and study areas, which span across the nation. To some degree, the exclusive use of national GIS datasets is a limitation because, for some datasets, the national data are less detailed in resolution and attribute information than similar GIS data for regional areas. The geographic extent of all references to "GIS thematic datasets" in this report is the conterminous United States.

The majority of the GIS thematic datasets used by NAWQA are either used in their native form or are slightly modified. The modifications are either changes in the projection and/or conversion from vector to raster format. The format and scale or resolution of the original and revised datasets that are used by NAWQA are included in the appendix.

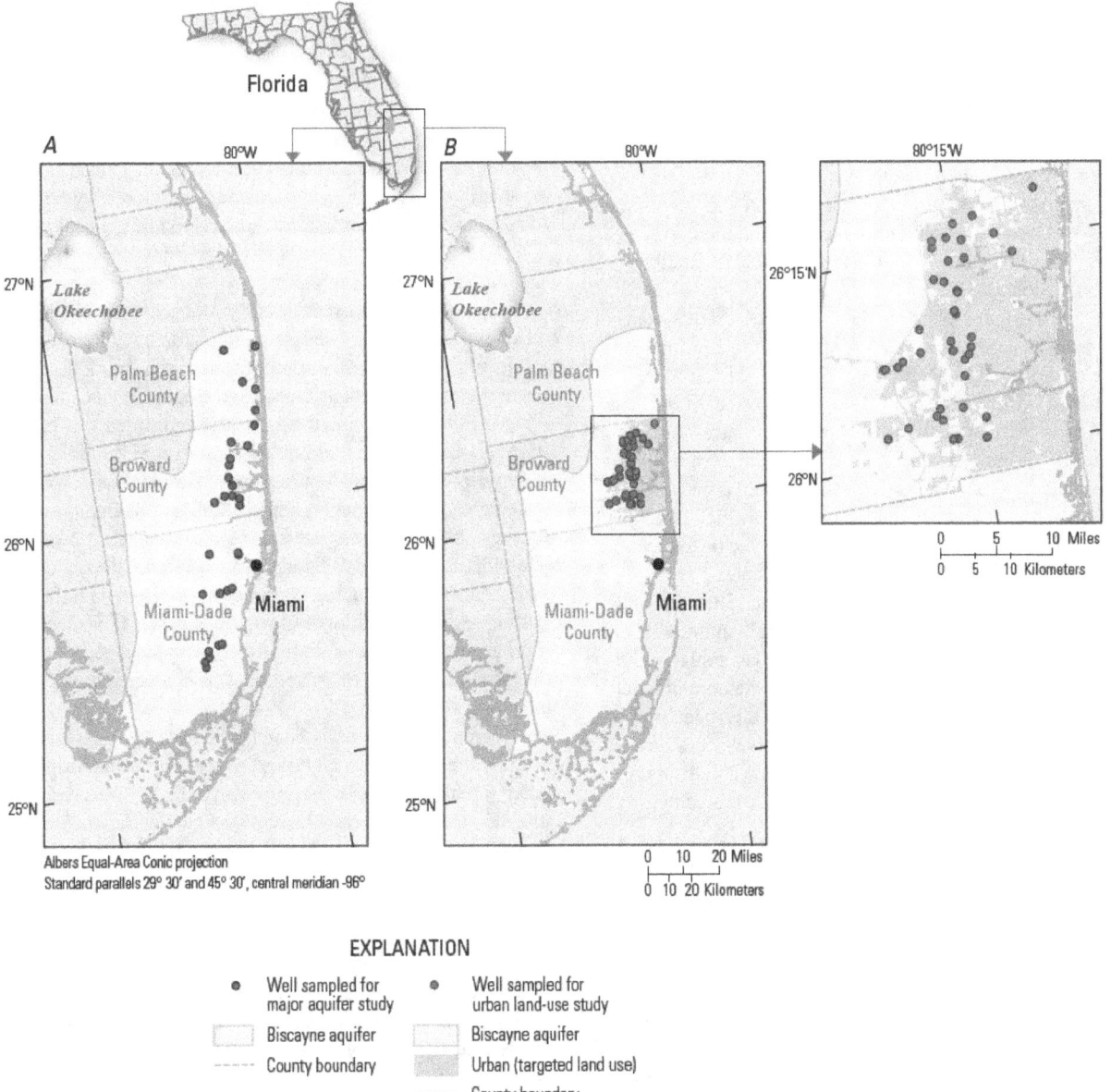

Figure 3. Geographic extent of and the location of wells (U.S. Geological Survey, 2010) sampled in (*A*) the major aquifer study of the Biscayne aquifer in southern Florida (David S. McCulloch, U.S. Geological Survey, written commun.,1998) and (*B*) the urban land-use study located within the Biscayne aquifer in southern Florida (David S. McCulloch, U.S. Geological Survey, written commun.,1998).

Methods

Geospatial datasets are composed of geographic and attribute information. The geographic component provides the location of the mapped features as point, line, or polygon features ("vector" data) or grid cells ("raster" data), and the attribute component provides characteristics of the mapped features. Mapped features can be represented as "continuous" data—such as elevation represented as continuous surfaces— or as "discrete" data—such as land cover represented by definable boundaries.

The attributes of mapped features are stored primarily in two ways: directly in the feature attribute table or indirectly in related data tables. With raster data, the grid cells are identified by their numeric values, which usually represent a single attribute, and the unique values are typically stored in the raster attribute table along with the number of grid cells represented by each value. With vector data, every mapped feature has an associated unique identifier, which is linked to its attributes. The attributes of mapped vector features are commonly stored in the feature attribute table.

For GIS datasets (vector or raster) that consist of a large number of attributes, however, it is generally more efficient to store the attributes in separate, related data tables, or, in other words, multiple tables each linked by a common field or attribute. Often, GIS data structured in this manner consist of mapped areal units that are administrative or political in nature. An example of mapped areal units with attributes stored in related data tables are counties attributed by population and number of housing units.

Multiple methods are used to characterize NAWQA sampling sites and associated study areas depending on the structure of the GIS thematic dataset. In this report, three characterization methods are described: (1) simple overlay, (2) area-weighted areal interpolation, and (3) land-cover-weighted areal interpolation. Simple overlay is used with GIS thematic datasets that store attribute information directly associated with each mapped feature (polygon, point, line, or grid cell), whereas areal interpolation is used with GIS thematic datasets in which attribute information is linked to mapped areal units. For all methods, the ArcInfo Workstation GRID commands that set the geoprocessing cell size and "snap grid" were applied to ensure all raster processing was done at 30-m resolution, with all cell locations aligned to the National Land Cover Dataset 1992 (U.S. Geological Survey, 1999; Vogelmann and others, 2001). The method that NAWQA used to process each GIS thematic dataset for NAWQA site and study-area characterization is noted in the appendix.

Simple Overlay

Simple overlay is the most basic method of characterizing a site or study area. The simple point overlay can be used to overlay (1) point locations, such as the sampling sites with areal features of interest like land cover, or (2) study-area boundaries with point features of interest like dams. The simple line overlay is used to overlay study-area boundaries with linear features of interest, such as roads, whereas the simple areal overlay is used to overlay study-area boundaries with areal features of interest.

Figure 4 shows examples of how a sampling site can be characterized by using simple point overlay. A single overlay of the national NAWQA sampling sites with a national thematic raster results in a table of the categorical or numeric value of the intersecting thematic grid cell for each site. Figure 4A shows the Pudding River at Aurora sampling site superimposed on the 30-m resolution National Land Cover Database 2001 (Homer and others, 2007; U.S. Geological Survey, 2007). The site intersects the land-cover grid cell classified as "woody wetlands," (fig. 4A), and, thus, the site is characterized as such. The same site overlain with the 100-m resolution National Elevation Dataset (U.S. Geological Survey, 2003), is shown in figure 4B. The intersecting grid cell has a numeric value of 26 and, thus, is characterized as having an elevation of 26 m.

Simple point overlay can also be used to characterize a study area by the density of point features of interest. The study-area boundary is overlain with the point features, and the results of this overlay provide the tabular data needed to calculate the total number of features within the study area. This number is then divided by the area of the study-area boundary to determine the feature's density within the study area.

Similarly, simple line overlay can be used to determine the density of linear features of interest within a study area. The process is nearly identical to acquiring the density of point features in a study area: (1) the study-area boundary is overlain with the linear (rather than point) features, (2) the total length (rather than number) of the linear features within the study area is calculated, and (3) the computed total length is divided by the study-area boundary to calculate the feature's density within the study area.

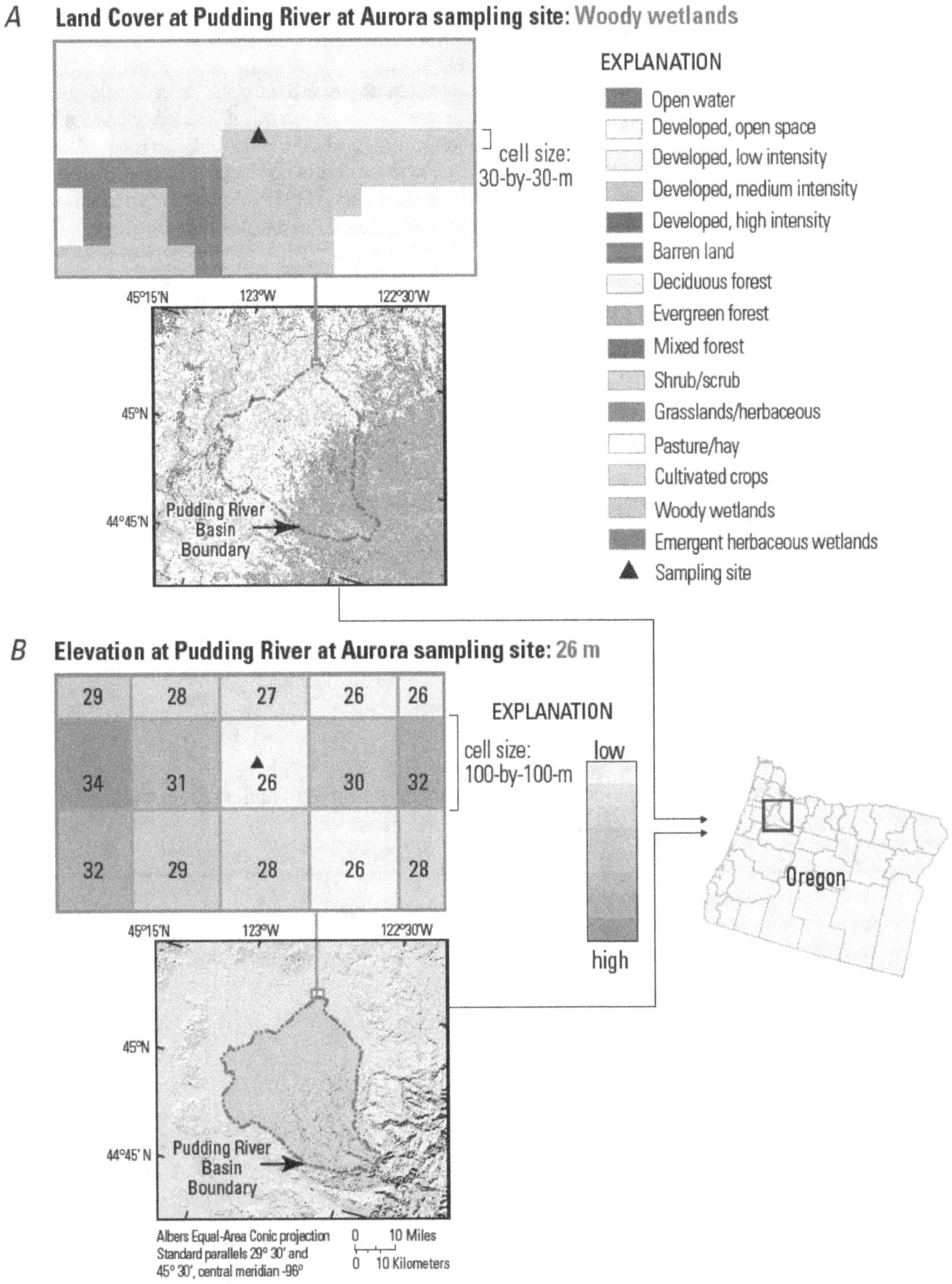

A   **Land Cover at Pudding River at Aurora sampling site:** Woody wetlands

B   **Elevation at Pudding River at Aurora sampling site:** 26 m

Figure 4.   Example showing the results of simple point overlay of the Pudding River sampling site (U.S. Geological Survey, 2010) with (*A*) a 30-m resolution land-cover raster from the National Land Cover Database 2001 (U.S. Geological Survey, 2007; Homer and others, 2007), and (*B*) a 100-m resolution elevation raster from the National Elevation Dataset (U.S. Geological Survey, 2003). m, meters.

To characterize a study area using simple areal overlay, the initial step is to overlay the study-area boundary with area-defined features, and the remaining steps are dependent upon whether the features are discrete or continuous. For discrete data, the study area is characterized as percentages or densities. In contrast, for continuous data, the study area is most often characterized as averages and rarely as total quantities or mass. Simple areal overlay using discrete and continuous data are illustrated in figures 5A and 5B, respectively.

The use of simple areal overlay to characterize land cover in the Pudding River basin is shown in figure 5A. The 30-m resolution land-cover raster is overlain with the 30-m resolution raster of the Pudding River basin boundary to generate a land-cover raster for the Pudding River basin. This land-cover raster includes the attribute table of land-cover classification codes (*Value*) and the number of (30-by-30-m resolution) grid cells (*Count*) for each land-cover classification in the Pudding River basin. The percentage of the land-cover classifications in the drainage basin is calculated by dividing the number of grid cells represented by each classification by the number of grid cells that define the basin boundary, then multiplying the quotient by 100. For example, the percentage of "open water" (land-cover code 11) for the Pudding River basin is the number of grid cells in the basin coded as 11 (7,500) divided by the number of grid cells that define the basin (1,500,000), multiplied by 100, which results in 0.5.

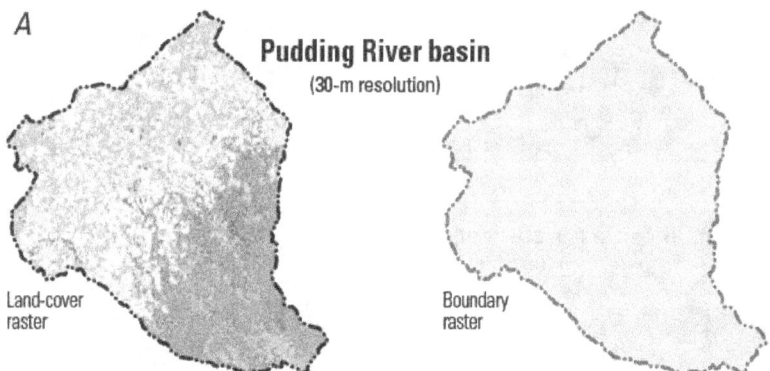

A

Pudding River basin
(30-m resolution)

Land-cover raster

Boundary raster

Attribute tables of the land-cover and boundary raster for the Pudding River basin, and derived percentages of land-cover classifications

Land-cover classification	Value	Count	Value	Count	Percentage land cover for basin
Open water	11	7,500	1	1,500,000	0.5
Developed, open space	21	30,000			2.0
Developed, low intensity	22	45,000			3.0
Developed, medium intensity	23	15,000			1.0
Developed, high intensity	24	7,500			0.5
Barren land	31	7,500			0.5
Deciduous forest	41	7,500			0.5
Evergreen forest	42	375,000			25.0
Mixed forest	43	15,000			1.0
Shrub/scrub	52	150,000			10.0
Grasslands/herbaceous	71	45,000			3.0
Pasture/hay	81	450,000			30.0
Cultivated crops	82	300,000			20.0
Woody wetlands	90	30,000			2.0
Emergent herbaceous wetlands	95	15,000			1.0

Figure 5.   Example showing the results of simple areal overlay of the Pudding River basin with (*A*) a 30-m resolution land-cover raster from the National Land Cover Database 2001 (U.S. Geological Survey, 2007; Homer and others, 2007) and (*B*) a 90-m resolution average annual precipitation raster. cm, centimeters; m, meters.

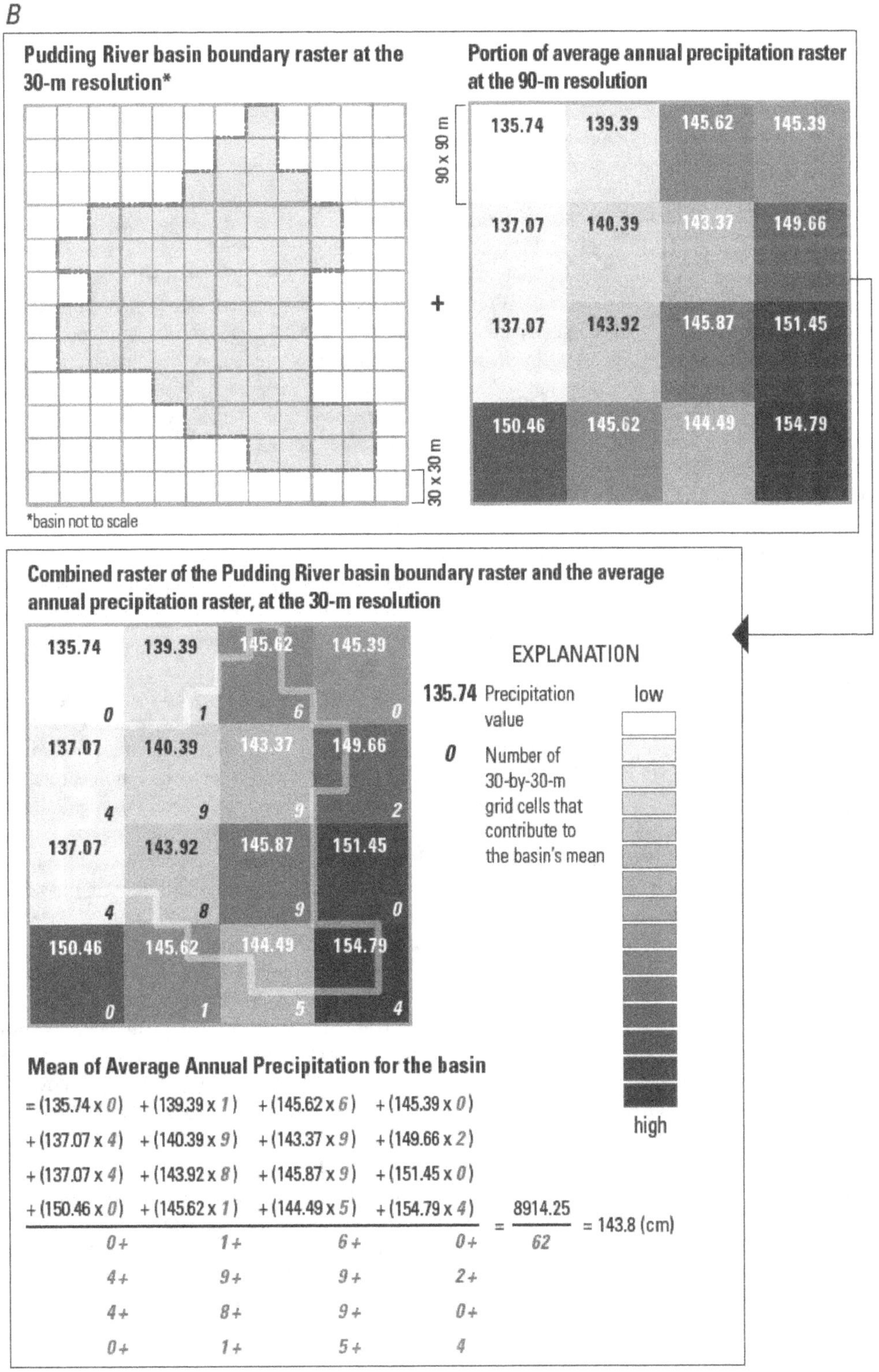

B

Pudding River basin boundary raster at the 30-m resolution*

*basin not to scale

Portion of average annual precipitation raster at the 90-m resolution

90 × 90 m

135.74	139.39	145.62	145.39
137.07	140.39	143.37	149.66
137.07	143.92	145.87	151.45
150.46	145.62	144.49	154.79

30 × 30 m

+

Combined raster of the Pudding River basin boundary raster and the average annual precipitation raster, at the 30-m resolution

135.74	139.39	145.62	145.39
0	*1*	*6*	*0*
137.07	140.39	143.37	149.66
4	*9*	*9*	*2*
137.07	143.92	145.87	151.45
4	*8*	*9*	*0*
150.46	145.62	144.49	154.79
0	*1*	*5*	*4*

EXPLANATION

135.74 Precipitation value

0 Number of 30-by-30-m grid cells that contribute to the basin's mean

low

high

Mean of Average Annual Precipitation for the basin

$$= \frac{\begin{array}{llll}(135.74 \times 0) & + (139.39 \times 1) & + (145.62 \times 6) & + (145.39 \times 0) \\ + (137.07 \times 4) & + (140.39 \times 9) & + (143.37 \times 9) & + (149.66 \times 2) \\ + (137.07 \times 4) & + (143.92 \times 8) & + (145.87 \times 9) & + (151.45 \times 0) \\ + (150.46 \times 0) & + (145.62 \times 1) & + (144.49 \times 5) & + (154.79 \times 4)\end{array}}{\begin{array}{llll} 0+ & 1+ & 6+ & 0+ \\ 4+ & 9+ & 9+ & 2+ \\ 4+ & 8+ & 9+ & 0+ \\ 0+ & 1+ & 5+ & 4 \end{array}} = \frac{8914.25}{62} = 143.8 \text{ (cm)}$$

Figure 5.—Continued

Figure 5B illustrates simple areal overlay using continuous data, and shows the computations that result when the study area and thematic data rasters have different grid-cell resolutions. In the example, a 90-m resolution (fictitious) precipitation raster is overlain with a 30-m resolution basin boundary raster, and the geoprocessing is set to 30-m. Geoprocessing the 90-m resolution precipitation raster at the 30-m resolution results in splitting or "resampling" each 90-by-90-m grid cell into nine 30-by-30-m grid cells, with each 30-by-30-m grid cell inheriting the value from the (undivided) 90-by-90-m grid cell. Note that only the 30-m resolution precipitation grid cells that overlap the 30-m resolution basin boundary, however, contribute to calculating the mean precipitation for the basin. The top left 90-by-90-m grid cell in figure 5B, with a value of 135.74, does not overlap the 30-m resolution basin boundary, so it does not contribute to computing the mean precipitation for the basin. The 90-by-90-m grid cell below the top left grid cell has four 30-by-30-m (resampled) precipitation grid cells that overlap the basin, so the values of these four grid cells (137.07) contribute to calculating the basin's mean precipitation. This spatial overlay analysis is repeated for each overlapping 30-by-30-m grid cell by the GIS software. The software computes the mean precipitation for the basin by taking the sum of the values of the 30-by-30-m precipitation grid cells that intersect the basin and dividing it by the total number of 30-by-30-m precipitation grid cells that were summed. The computation of the mean average annual precipitation for the Pudding River basin, shown in figure 5B, is 8914.25 divided by 62, or 143.8 (centimeters).

The NAWQA Program processes many 1-km resolution national GIS thematic datasets to calculate study-area means of various features. When the resolution of the thematic dataset (for example, 1-km or 1000-m) does not divide evenly into the processing resolution (such as 30-m), an uneven resampling of grid cells takes place. In ArcInfo GRID, "nearest-neighbor resampling" occurs by default (which is the setting applied by NAWQA), and the 30-by-30-m resampled thematic data grid cells that partially intersect the basin or study area contribute to its computed mean only if the centroid of these grid cells intersects the 30-by-30-m grid cell of the study area in the same location.

Areal Interpolation

A common geospatial problem is estimating quantities of an attribute that are defined by mapped areal units that have different boundaries than the study areas of interest. One can acquire crude estimates for the study area by summing the quantities of the attribute corresponding to the mapped areal units, or "source zones," that entirely or partially intersect the study area, or "target zone." While the inclusion of the quantities in their entirety for the partially intersecting source zones would likely produce an overestimation, the exclusion of the quantities for these subdivided source zones would likely produce an underestimation. Instead, the quantities could be refined for the subdivided source zones if areal interpolation were applied to proportion the quantities by the extent of intersection of the source and target zones. In this report, this method is referred to as "area-weighted" areal interpolation.

A simplified example of estimating quantities using area-weighted areal interpolation is shown in figure 6A. There are two fictitious source zones, "X" and "Y," which have a quantity of 6 and 15 units, respectively. Both source zones intersect the target zone (the dotted rectangle), and the quantity that each of these intersecting source zones contributes to the target zone's estimate is dependent upon the fraction of the source zone that intersects the target zone. In this example, the amount of overlap of source zone X with the target zone is one-sixth, so one-sixth of the quantity associated with source zone X (6 units) contributes to the target zone (1 unit). About one-third of source zone Y intersects the target zone, so one-third of 15, or 5 units of source zone Y contribute to the total quantity for the target zone. The area-weighted total quantity for the target zone then is computed by taking the sum of the contributions from the two source zones (1 and 5 units), which is 6 units. Area-weighted areal interpolation assumes that there is homogeneous distribution throughout the source zones.

An approach to improving area-weighted areal interpolation is to use mapped land cover to refine the spatial distribution of the quantified attribute within the source zone. The selected land-cover classification is related to the attribute of interest, for example, mapped residential urban areas and population. Therefore, with land-cover weighted areal interpolation, the spatial distribution of the quantified attribute is confined only to areas that represent the selected land-cover classification. The distribution within these areas is also assumed to be homogeneous.

An example of estimating quantities using land-cover weighted areal interpolation is shown in figure 6B. The red-colored boxes represent the mapped areas of a specific land-cover classification. The same quantities for source zones X and Y depicted in figure 6A are now confined to the red boxes in the source zones, such that the red area in source zones X and Y has a quantity of 6 and 15 units, respectively (7.5 for each red box in source zone Y). One-quarter of the area of the selected land-cover classification in source zone X intersects with the target zone; therefore, one-fourth of the quantity associated with source zone X (one-quarter of 6, or 1.5 units) contributes to the total estimate for the target zone. Because the two areas of the selected land-cover classification in the source zone Y do not intersect the target zone, the quantities associated with source zone Y do not contribute to the target zone's estimate. The total quantity for the basin using land-cover weighted interpolation is the sum of the land-cover weighted contributions from source zones X and Y (1.5 + 0, or 1.5 units).

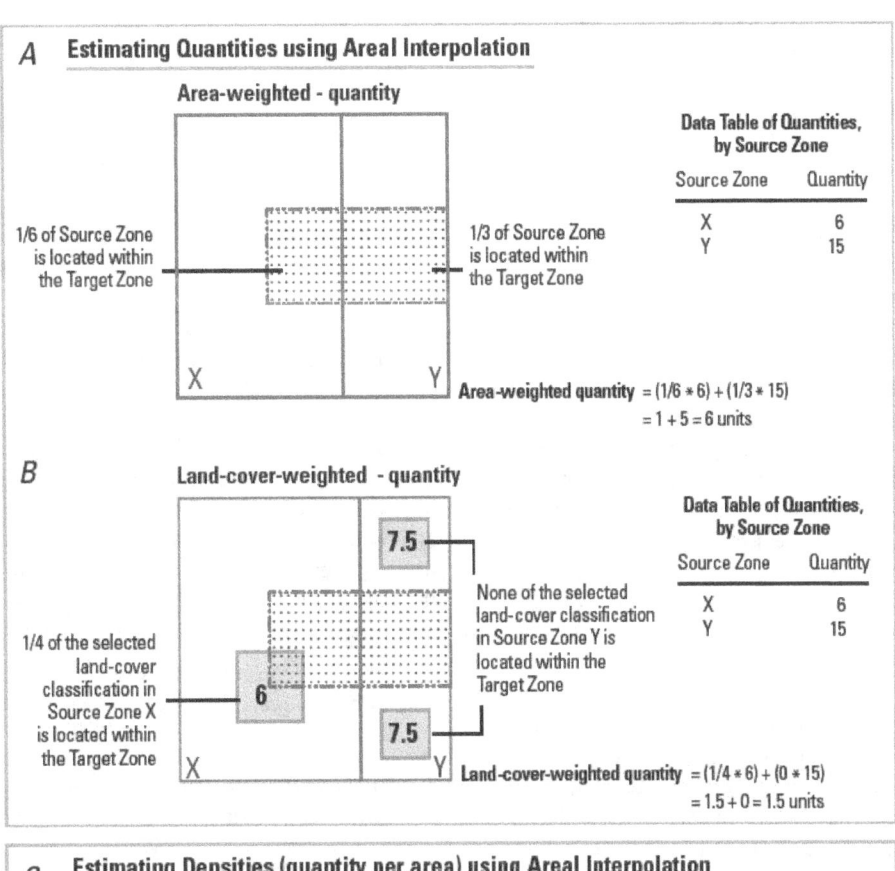

A   **Estimating Quantities using Areal Interpolation**

Area-weighted - quantity

1/6 of Source Zone is located within the Target Zone

1/3 of Source Zone is located within the Target Zone

Data Table of Quantities, by Source Zone

Source Zone	Quantity
X	6
Y	15

Area-weighted quantity $= (1/6 * 6) + (1/3 * 15)$
$= 1 + 5 = 6$ units

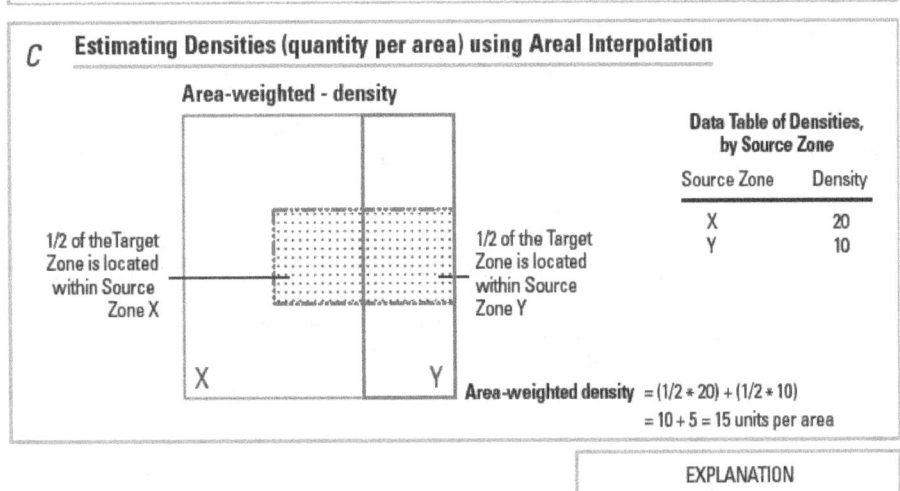

B   **Land-cover-weighted - quantity**

7.5

6

7.5

1/4 of the selected land-cover classification in Source Zone X is located within the Target Zone

None of the selected land-cover classification in Source Zone Y is located within the Target Zone

Data Table of Quantities, by Source Zone

Source Zone	Quantity
X	6
Y	15

Land-cover-weighted quantity $= (1/4 * 6) + (0 * 15)$
$= 1.5 + 0 = 1.5$ units

C   **Estimating Densities (quantity per area) using Areal Interpolation**

Area-weighted - density

1/2 of the Target Zone is located within Source Zone X

1/2 of the Target Zone is located within Source Zone Y

Data Table of Densities, by Source Zone

Source Zone	Density
X	20
Y	10

Area-weighted density $= (1/2 * 20) + (1/2 * 10)$
$= 10 + 5 = 15$ units per area

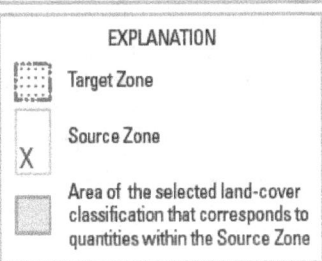

EXPLANATION

Target Zone

Source Zone

Area of the selected land-cover classification that corresponds to quantities within the Source Zone

Figure 6.   Simplified examples of estimating quantities and densities by using areal interpolation.

Although the examples in figures 6A and 6B are fictitious, they show how estimates of quantities can be significantly different depending on the method used to obtain them (by a factor of four). Note that interpolation is unnecessary for a source zone that is entirely within the target zone with either area-weighted or land-cover-weighted areal interpolation because the entire quantity associated with the undivided source zones contributes to the total for the study area. With areal interpolation, in general, the greater amount of overlap that exists between the target and source zones, the higher the probability for more accurate weighted estimates for the subdivided source zones.

For estimating densities, or quantities per unit area, such as mean population density, the intersecting source zone's contribution is based on the percentage of the target zone intersecting each source zone. The percentage is applied to the density associated with the entire source zone to quantify the contribution from each intersecting source zone. Thus, if 80 percent of the target zone overlaps a source zone, then 80 percent of the density associated with this source zone contributes to the density of the target zone. This approach also assumes the density of the attribute is uniform across the source zones.

An example of applying area-weighted interpolation to determine an area-weighted mean density is depicted in figure 6C. As with the examples in figures 6A and 6B, source zones X and Y intersect the target zone; however, in figure 6C, 20 and 10 are the associated attribute densities for the two source zones, respectively. Because half of the target zone lies in each of the two source zones, half of the density associated with source zone X (20) contributes to the weighted mean for the target zone (0.5 multiplied by 20, or 10) and half of the density associated with source zone Y (10) contributes to the weighted mean for the target zone (0.5 multiplied by 10, or 5). The sum of the weighted means (10 and 5, respectively), or 15 units per area, is the resulting weighted mean density for the target zone.

The following two sections walk through step-by-step calculations for estimating mean densities and total quantities of an attribute for a study area using area-weighted areal interpolation, and estimating total quantities (or masses) of an attribute derived from land-cover weighted areal interpolation. Hereafter, the estimates derived for total quantities and mean densities using area-weighted areal interpolation are referred to as "area-weighted quantities" and "area-weighted mean densities," respectively; the estimates computed for quantities using land-cover weighted interpolation are referred to as "land-cover weighted quantities."

Area-Weighted Areal Interpolation

Area-weighted areal interpolation is used to estimate mean densities and quantities of an attribute. The process of estimating the mean density of an attribute for a target zone (Dt) can be summarized as follows:

$$\sum_{i=1}^{n}\left(\frac{Ast_i}{At} * Ds_i\right) \quad\quad (1)$$

where

n is number of source zones (for example, counties) intersecting the target zone (for example, drainage basin),

Ast_i is Area of intersection, of source zone i and the target zone,

At is Area of the target zone, and

Ds_i is Density of an attribute expressed in units per area, for source zone i.

The use of area-weighted areal interpolation to estimate the area-weighted mean (or the weighted average) population density for the Pudding River basin using county boundaries and (fictitious) population density values is illustrated in figure 7. Here, the target zone is the Pudding River basin and the source zones are counties, for which there are population density statistics. The method is presented in 4 steps (fig. 7):

1. determine the areas of intersection of the source zones with the target zone (Ast), and determine the area of the target zone (At);

2. calculate the area-weighted weighting factors for the source zones that intersect the target zone (WF);

3. calculate the area-weighted densities of the attribute for the source zones that intersect the target zone (Dst); and

4. sum the area-weighted densities (Dst) to calculate the area-weighted mean density of the attribute for the target zone (Dt).

Step 1 has two parts. The first part of step 1 (fig. 7, step 1A) is to determine the areas of intersection of the source zones (counties) with the target zone (Pudding River basin). To compute these areas, the raster of the Pudding River basin boundary is overlain with the national raster of county boundaries. The results from this raster overlay yield the number of grid cells (and thus the areas) of the Pudding River basin that intersect each county. In this example, the basin intersects two counties, Clackamas and Marion (referred to as counties "A" and "B," respectively), and the areas of the counties intersecting the basin are 300 and 900 km², respectively. The second part of step 1 (fig. 7, step 1B) is to determine the area of the target zone. Derived from the number of grid cells that define the basin boundary, the area of the Pudding River basin is shown to be 1,200 km².

Next, step 2 (fig. 7, step 2) calculates the area-weighted weighting factors for the source zones that intersect the target zone. This weighting factor is equal to the area of intersection of the source zone with the target zone, divided by the area of the target zone. In this example, the weighting factor for county A is 300 km² divided by 1200 km², or 0.25; for county B, it is 900 km² divided by 1,200 km², or 0.75.

Step 1. **a)** Determine the areas of intersection of the source zones (counties) with the target zone (Pudding River basin)—(Ast).

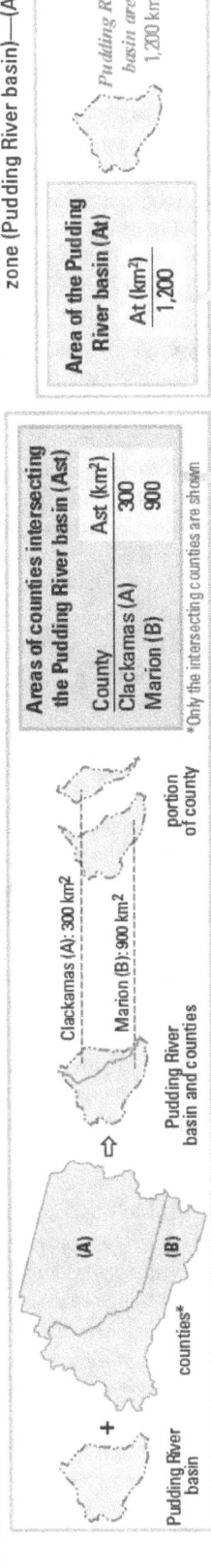

Clackamas (A): 300 km²

Marion (B): 900 km²

portion of county

Pudding River basin and counties

Pudding River basin   counties*

Step 1. **b)** Determine the area of the target zone (Pudding River basin)—(At).

Pudding River basin area: 1,200 km²

Area of the Pudding River basin (At)
At (km²)
1,200

Areas of counties intersecting the Pudding River basin (Ast)	
County	Ast (km²)
Clackamas (A)	300
Marion (B)	900

*Only the intersecting counties are shown

Step 2. Calculate the weighting factors for the source zones that intersect the target zone—(WF).

$$WF_A = \left(\frac{Ast_A}{At}\right) = \frac{300}{1,200} = 0.25$$

$$WF_B = \left(\frac{Ast_B}{At}\right) = \frac{900}{1,200} = 0.75$$

Weighting factors (WF) by county, areas of counties intersecting the Pudding River basin (Ast), and area of the Pudding River basin (At)

County	WF	Ast (km²)	At (km²)
Clackamas (A)	0.25	300	1,200
Marion (B)	0.75	900	1,200

Step 3. Calculate the area-weighted densities (of population) for the source zones that intersect the target zone—(Dst).

$$Dst_A = (WF_A * Ds_A)$$
$$= 0.25 * 80 \text{ persons/km}^2$$
$$= 20 \text{ persons/km}^2$$

$$Dst_B = (WF_B * Ds_B)$$
$$= 0.75 * 60 \text{ persons/km}^2$$
$$= 45 \text{ persons/km}^2$$

Population densities (Ds), countywide	
County	Ds (persons/km²)
Clackamas (A)	80
Marion (B)	60
. . .	. . .

Area-weighted population densities (Dst) by county, weighting factors (WF) by county, and population densities (Ds), countywide

County	Dst (persons/km²)	WF	Ds (persons/km²)
Clackamas (A)	20	0.25	80
Marion (B)	45	0.75	60

Step 4. Sum the area-weighted densities (of population) to calculate the estimated mean density for the target zone (Pudding River basin)—(Dt).

$$Dt = Dst_A + Dst_B = 20 + 45 = 65 \text{ persons/km}^2$$

or

$$Dt = \left(\frac{Ast_A}{At} * Ds_A\right) + \left(\frac{Ast_B}{At} * Ds_B\right)$$

Estimated mean population density (Dt) for the Pudding River basin
Dt (persons/km²)
65

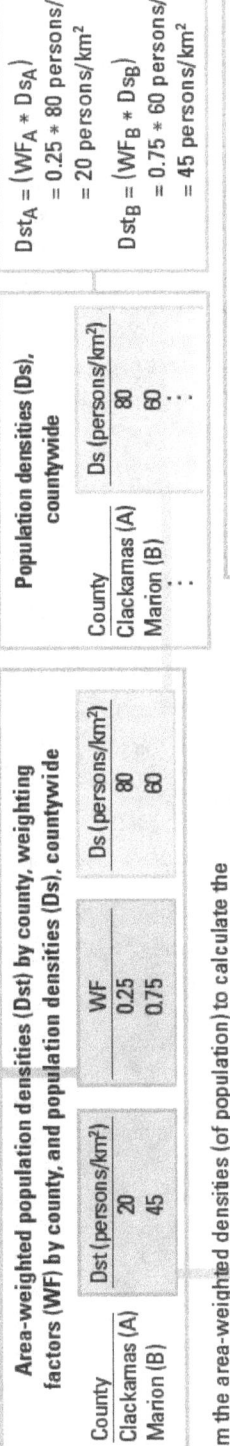

Clackamas County (A) population density: 80

Marion County (B) population density: 60

County boundaries raster and the Pudding River basin boundary raster

Clackamas County (A): Area-weighted population density: 20

Marion County (B): Area-weighted population density: 45

Figure 7. Example of area-weighted areal interpolation to estimate mean population density for the Pudding River basin by using (1) a raster of the Pudding River basin boundary, (2) a raster of county boundaries, and (3) fictitious population density statistics by county. The county boundaries and basin boundary are the source zones and target zone, respectively. "Ast" stores the area of counties intersecting the basin; "At" stores the area of the basin; "WF" stores the area-weighted weighting factors for estimating densities; "Ds" stores the population density by county; "Dst" stores the area-weighted population densities by county; and, "Dt" stores the area-weighted mean population density for the Pudding River basin. km², square kilometer.

In step 3 (fig. 7, step 3), the area-weighted density of the attribute (population) is determined for the source zones that intersect the target zone by multiplying the weighting factor for the source zone by the density for the source zone. Given that the countywide population densities for counties A and B are 80 and 60 persons/km², respectively, the weighted population density estimates for the portions of these two counties in the Pudding River basin are 0.25 multiplied by 80 persons/km², or 20 persons/km², and 0.75 multiplied by 60 persons/km², or 45 persons/km², respectively.

Finally, in step 4 (fig. 7, step 4), the estimated area-weighted density of the attribute (population) for the target zone (Pudding River basin) is computed by summing the weighted densities (of population) calculated in the previous step. The sum of the weighted densities for counties A and B (20 and 45 persons/km²) is 65 persons/km².

The process of estimating quantities of an attribute for a target zone by using area-weighted areal interpolation (Qt) can be summarized as follows:

$$\sum_{i=1}^{n}\left(\frac{Ast_i}{As_i}*Qs_i\right)\qquad(2)$$

where

    n is number of source zones (for example, counties) intersecting the target zone (for example, drainage basin),

    Ast_i is Area of intersection, of source zone i with the target zone,

    As_i is Area of the source zone i, and

    Qs_i is Quantity of an attribute for source zone i.

Equations 1 and 2 have important differences: the divisor in equation 1 represents the area of the target zone, whereas the divisor in formula 2 represents the area of the source zone. In other words, the only variation lies in the computation of the weighting factor. Accordingly, the steps to compute area-weighted quantities are nearly identical to the steps to compute area-weighted mean densities. The step-by-step process to estimate population for the Pudding River basin using area-weighted areal interpolation from county boundaries and (fictitious) county areas and population statistics is illustrated in figure 8.

The initial step in this approach determines the areas of intersection of the source zones with the target zone (fig. 8, step 1A), which is identical to figure 7 step 1A; however, the second part (fig. 8, step 1B determines the area of the source zones (counties) rather than the target zone. Derived from the number of grid cells for each county in the national raster of county boundaries, the areas of Clackamas and Marion counties (counties "A" and "B," respectively) are 5,000 km² and 3,000 km², respectively. (Note: this step only needs to be carried out a single time to obtain the source-zone areas for the entire nation.)

Next, step 2 (fig. 8, step 2) calculates the weighting factors for estimating quantities for the source zones (counties) that intersect the target zone. The weighting factor is equal to the area of intersection of the source zone with the target zone, divided by the total area of the source zone. In this example, the weighting factor for county A is 300 km² divided by the countywide area, 5,000 km², or 0.06; and, similarly, the weighting factor for county B is 900 km² divided 3,000 km², or 0.30.

In step 3 (fig. 8, step 3), the weighted quantity of the attribute (population) for a source zone that intersects the target zone is determined by multiplying the weighting factor for a source zone by the quantity (population) for the source zone (county). Given the populations for counties A and B are 250,000 and 210,000 persons, respectively, the weighted populations for these two counties are 0.06 multiplied by 250,000 persons, or 15,000 persons, and 0.3 multiplied by 210,000 persons, or 63,000 persons, respectively.

Finally, in step 4 (fig. 8, step 4), the quantitative estimate of the attribute (population) for the target zone (Pudding River basin) is computed by summing the weighted quantities of the attribute (population) calculated in the previous step. The weighted population for counties A and B, 15,000 and 63,000 persons, respectively, are summed, which is 78,000 persons. Population density could then be calculated by dividing the 78,000 persons by the area of the basin, 1,200 km², which is 65 persons/km².

It is worth noting that if a national raster of population density or population counts was readily available, simple overlay could be used instead of areal interpolation. The population density raster could be superimposed over a basin boundary raster, for example, and the mean of the population density values of the overlapping grid cells would reflect the mean population density for the basin. Similarly, a population (counts) raster could be superimposed over the basin boundary raster, and the sum of the population values of the overlapping grid cells would reflect the total population for the basin. In addition, this total population estimate divided by the basin area would represent the mean population density for the basin.

Two factors are important to consider when deciding whether to use simple overlay or areal interpolation if the option exists: the resolution of the raster data for the attribute of interest, and the size of the source and target zones. A raster of population density or population counts with large grid cells (for example, 5 kilometers by 5 kilometers) would be too coarse to characterize a number of small drainage basins (for example, less than 25 km²). Results could be improved by applying area-weighted interpolation using detailed source-zone boundaries rasterized at a higher resolution (smaller grid cells). Furthermore, estimating population density or population (counts) using smaller source zones (such as census blocks rather than counties) would greatly improve estimates for the partially intersecting source zones. If, however, either method could be applied, but there were a large number of source-zone level attributes of interest, it would be more efficient to employ areal interpolation instead of simple overlay to obtain study-area characteristics.

Step 1. **a)** Determine the areas of intersection of the source zones (counties) with the target zone (Pudding River basin)—(Ast).   **Step 1.** **b)** Determine the areas of the source zones (counties)—(As).

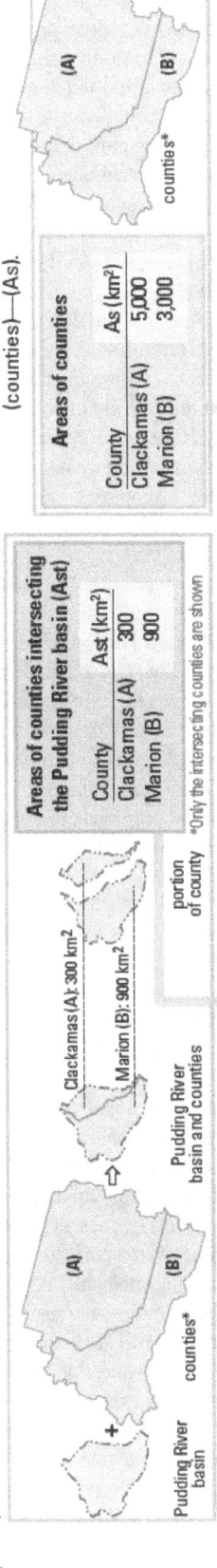

Areas of counties intersecting the Pudding River basin (Ast)

County	Ast (km²)
Clackamas (A)	300
Marion (B)	900

*Only the intersecting counties are shown

Areas of counties

County	As (km²)
Clackamas (A)	5,000
Marion (B)	3,000

Step 2. Calculate the weighting factors for the source zones that intersect the target zone—(WF).

Weighting factors (WF) by county, areas of counties intersecting the Pudding River basin (Ast), and areas of counties (As)

County	WF	Ast (km²)	As (km²)
Clackamas (A)	0.06	300	5,000
Marion (B)	0.30	900	3,000

$$WF_A = \left(\frac{Ast_A}{As_A}\right) = \frac{300}{5,000} = 0.06$$

$$WF_B = \left(\frac{Ast_B}{As_B}\right) = \frac{900}{3,000} = 0.30$$

Step 3. Calculate the area-weighted quantities (of persons) for the source zones that intersect the target zone—(Qst).

Area-weighted persons (Qst) by county, weighting factors (WF) by county, and persons (Qs), countywide

County	Qst (persons)	WF	Qs (persons)
Clackamas (A)	15,000	0.06	250,000
Marion (B)	63,000	0.30	210,000

Persons (Qs), countywide

County	Qs (persons)
Clackamas (A)	250,000
Marion (B)	210,000
...	...

$$Qst_A = (WF_A * Qs_A)$$
$$= 0.06 * 250,000$$
$$= 15,000 \text{ persons}$$

$$Qst_B = (WF_B * Qs_B)$$
$$= 0.30 * 210,000$$
$$= 63,000 \text{ persons}$$

Step 4. Sum the area-weighted quantities (of persons) to calculate the estimated quantity for the target zone (Pudding River basin)—(Qt).

Estimated persons for the Pudding River basin

Qt (persons)
78,000

$$Qt = Qst_A + Qst_B = 15,000 + 63,000 = 78,000 \text{ persons}$$

or

$$Qt = \left(\frac{Ast_A}{As_A} * Qs_A\right) + \left(\frac{Ast_B}{As_B} * Qs_B\right)$$

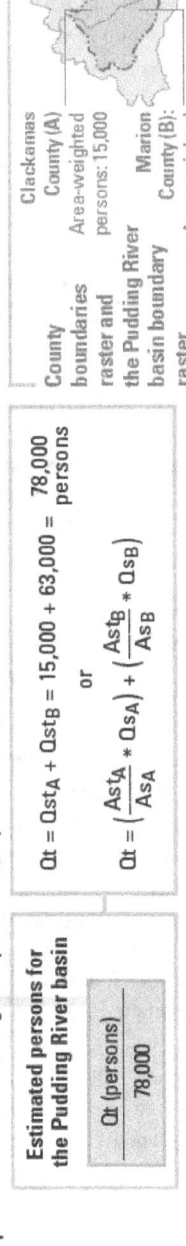

County boundaries raster and the Pudding River basin boundary raster

Clackamas County (A) Area-weighted persons: 15,000

Marion County (B): Area-weighted persons: 63,000

Clackamas County (A) Persons: 250,000

Marion County (B) Persons: 210,000

Figure 8. Example of area-weighted areal interpolation to estimate population for the Pudding River basin by using (1) a national raster of county boundaries, (2) a raster of the Pudding River basin boundary, and, (3) fictitious population statistics by county. The county boundaries and basin boundary are the source zones and target zone, respectively. "*Ast*" stores the area of a counties intersecting the basin; "*As*" stores the area of the county; "*WF*" stores the area-weighted weighting factor for estimating quantities; "*Qs*" stores the total population by county; "*Qst*" stores the area-weighted population by county; and, "*Qt*" stores the area-weighted sum of population for the Pudding River basin. km², square kilometer.

With areal interpolation, in general, the map overlay, which is the most computationally-intensive step, is conducted once to determine the weighting factors, and the remaining processing consists of simple tabular computations of any number of variables. If, for example, characteristics for drainage basins located throughout the U.S. were needed for dozens of census variables, simple overlay would require a map overlay of the basin boundary with each national raster of the census variables. The use of areal interpolation instead of simple overlay for characterizing numerous variables by source zones reduces processing time and avoids data duplication.

Land-Cover-Weighted Areal Interpolation

Land-cover-weighted areal interpolation is an extension of area-weighted areal interpolation in which land cover is added to refine the spatial interpolation process. Rather than distributing a quantified attribute uniformly throughout the source zone, the attribute is confined to areas representing selected land classifications that have a direct association with the attribute. The addition of ancillary data for areal interpolation, which has been referred to as "dasymetric mapping," has been used to estimate population using census geography and urban land cover from classified satellite imagery (Riebel and Agrawal, 2007; Holt and others, 2004; Fisher and Langford, 1996).

Although any single land classification or groupings of classifications can be used with land-cover-weighted areal interpolation, the following discussion and example illustrate the method applied to cropland. The process of estimating quantities of an attribute for a target zone by using land-cover-weighted areal interpolation (QLt) can be summarized as follows:

$$\sum_{i=1}^{n} \left(\frac{ALst_i}{ALs_i} * Qs_i \right)$$ (3)

where

n is number of source zones (for example, counties) intersecting a target zone (for example, drainage basin),

$ALst_i$ is Area of intersection, of the selected land-cover classification (for example, cropland) within source zone i and the target zone,

ALs_i is Area of the selected land-cover classification in the entire source zone i, and

Qs_i is Quantity of an attribute for source zone i.

Equations 2 and 3 have one difference: in equation 2, the areas (Ast_i and As_i) pertain to the source zones in general and, in equation 3, the areas ($ALst_i$ and ALs_i) pertain to the areas of land-cover classifications within the source zones.

The implementation of land-cover-weighted areal interpolation is illustrated in figure 9. The example shows how it can be used to estimate the amount of atrazine applied on cultivated crops in the Pudding River basin by using (fictitious) atrazine-use values and areas of cropland. In this example, the Pudding River basin is the target zone, the counties are the source zones, estimated amount of atrazine use on crops by county is the quantitative attribute for the source zones, and "cultivated crops" is the selected land-cover classification. The source for mapped cropland is the National Land Cover Database 2001 or "NLCD 2001" (U.S. Geological Survey, 2007; Homer and others, 2007).

The first part of step 1 (fig. 9, step 1A) is to determine, by source zone (county), the total area of the selected land classification, "cultivated crops," located within the target zone (Pudding River basin). These areas are obtained by overlaying the raster of the Pudding River basin boundary with the national rasters of land cover and county boundaries. The results from these overlays yield the county areas of cropland in the basin: for Clackamas county (county "A"), the total area of land classified as "cultivated crops" within the basin is the 75 km^2; for Marion county (county "B"), the total area is 200 km^2.

The second part of step 1 (fig. 9, step 1B) is to determine the source-zone (countywide) areas of the selected land classification ("cultivated crops"). These areas are acquired by overlaying the national land-cover raster with the national county boundaries raster. The countywide area of "cultivated crops" for counties A and B are 300 km^2 and 500 km^2, respectively. (Note: this step is required only once to obtain the source-zone areas of the selected land cover for the entire nation.)

Figure 9.   Example of land-cover-weighted areal interpolation to estimate amount of atrazine applied on cultivated crops for the Pudding River basin by using (1) a 30-m resolution land-cover raster from the National Land Cover Database 2001 or "NLCD 2001" (U.S. Geological Survey, 2007; Homer and others, 2007), (2) a national raster of county boundaries, (3) a raster of the Pudding River basin boundary, and (4) fictitious county atrazine use estimates on cultivated crops. The county boundaries and basin boundary are the source zones and target zone, respectively. "ALst" stores the area of land classified as "cultivated crops" in the NLCD 2001 located within the Pudding River basin, by county; "ALs" stores the countywide area of land classified as "cultivated crops" in the NLCD 2001, by county; "WF" stores the land-cover weighted weighting factor for quantifying totals; "Qs" stores the atrazine use value for the county; "Qst" stores the weighted value of atrazine for the county; and "Qt" stores the land-cover-weighted total mass of atrazine use for the Pudding River basin. km^2, square kilometer; kg, kilograms. Shown on next page.

Step 1 a) Determine by source zone (county), the total areas of the selected land classification ("Cultivated crops") located within the target zone (Pudding River basin)—(ALst).

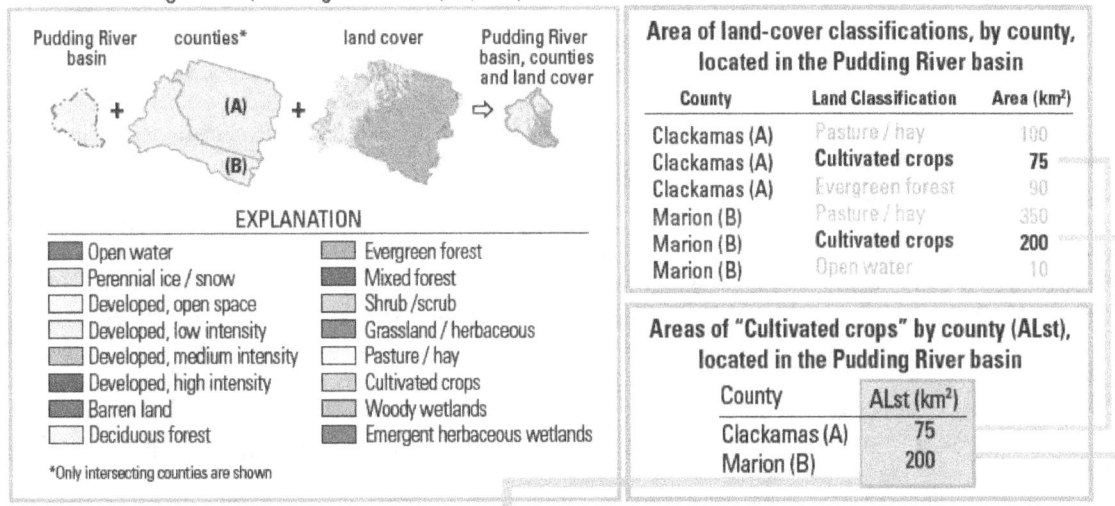

Area of land-cover classifications, by county, located in the Pudding River basin

County	Land Classification	Area (km²)
Clackamas (A)	Pasture / hay	100
Clackamas (A)	**Cultivated crops**	**75**
Clackamas (A)	Evergreen forest	90
Marion (B)	Pasture / hay	350
Marion (B)	**Cultivated crops**	**200**
Marion (B)	Open water	10

Areas of "Cultivated crops" by county (ALst), located in the Pudding River basin

County	ALst (km²)
Clackamas (A)	75
Marion (B)	200

EXPLANATION

- Open water
- Perennial ice / snow
- Developed, open space
- Developed, low intensity
- Developed, medium intensity
- Developed, high intensity
- Barren land
- Deciduous forest
- Evergreen forest
- Mixed forest
- Shrub /scrub
- Grassland / herbaceous
- Pasture / hay
- Cultivated crops
- Woody wetlands
- Emergent herbaceous wetlands

*Only intersecting counties are shown

Step 1 b) Determine the source-zone-level (countywide) areas of the selected land classification ("Cultivated crops")—(ALs).

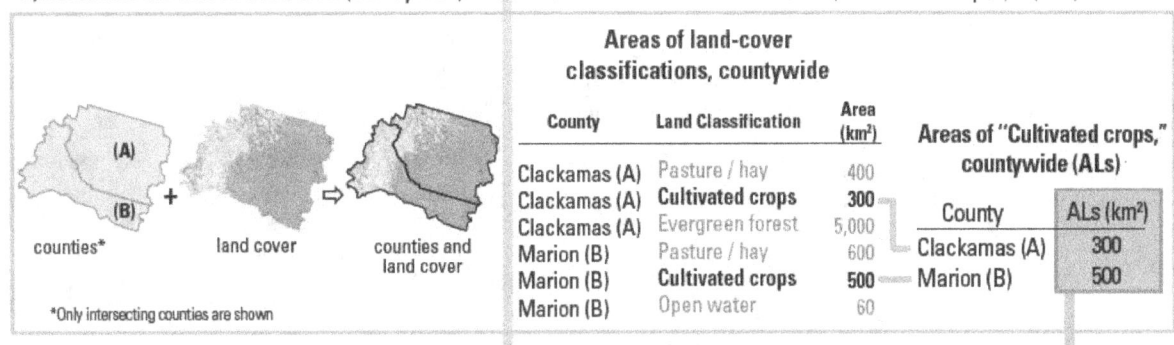

Areas of land-cover classifications, countywide

County	Land Classification	Area (km²)
Clackamas (A)	Pasture / hay	400
Clackamas (A)	**Cultivated crops**	**300**
Clackamas (A)	Evergreen forest	5,000
Marion (B)	Pasture / hay	600
Marion (B)	**Cultivated crops**	**500**
Marion (B)	Open water	60

Areas of "Cultivated crops," countywide (ALs)

County	ALs (km²)
Clackamas (A)	300
Marion (B)	500

counties*   +   land cover   ⇒   counties and land cover

*Only intersecting counties are shown

Step 2 Calculate the weighting factors for the source zones that intersect the target zones—(WF).

Weighting factors (WF), areas of "Cultivated crops" within the Pudding River basin (ALst), and areas of "Cultivated crops," countywide (ALs)

County	WF	ALst (km²)	ALs (km²)
Clackamas (A)	0.25	75	300
Marion (B)	0.40	200	500

$$WF_A = \left(\frac{ALst_A}{ALs_A}\right) = \frac{75}{300} = 0.25$$

$$WF_B = \left(\frac{ALst_B}{ALs_B}\right) = \frac{200}{500} = 0.40$$

Step 3 Calculate the land-cover weighted estimates (of atrazine use on crops) for the source zones that intersect the target zone—(Qst).

Weighted atrazine use (Qst), weighting factors (WF) and atrazine use on crops, countywide (Qs)

County	Qst (kg)	WF	Qs (kg)
Clackamas (A)	300	0.25	1,200
Marion (B)	3,200	0.40	8,000

Atrazine use (kg) on crops, countywide (Qs)

County	Qs (kg)
Clackamas (A)	1,200
Marion (B)	8,000
⋮	⋮

$$Qst_A = (WF_A * Qs_A)$$
$$= 0.025 * 1,200 \,(kg) = 300 \,kg$$

$$Qst_B = (WF_B * Qs_B)$$
$$= 0.040 * 8,000 \,(kg) = 3,200 \,kg$$

Step 4 Sum the weighted estimates to calculate the estimated amount (of atrazine use on crops) for the target zone (Pudding River basin)—(QLt).

Estimated amount of atrazine use on crops for the Pudding River basin (QLt)

QLt (kg)
3,500 kg

$$QLt = Qst_A + Qst_B = 300 + 3,200 = 3,500 \,kg$$

or

$$QLt = \left(\frac{ALst_A}{ALs_A} * Qs_A\right) + \left(\frac{ALst_B}{ALs_B} * Qs_B\right)$$

Then, step 2 (fig. 9, step 2) calculates the weighting factors for the selected land-cover classification for the source zones (counties) that intersect the target zone (Pudding River basin). In this example, the (cropland-weighted) weighting factor for a county is equal to the area of cultivated crops in the county and basin divided by the countywide area of cultivated crops. Accordingly, the weighting factors for counties A and B are 75 divided by 300, or 0.25, and 200 divided by 500, or 0.4, respectively.

In figure 9, step 3, the weighted attribute (of atrazine use) values for the source zones that intersect the target zone (Pudding River basin) are calculated. The weighted atrazine use value for a county is calculated by multiplying the county's weighting factor by the county's atrazine use value. The example shows 1,200 kg of atrazine is applied in county A and 8,000 kg in county B; thus, the weighted atrazine use for counties A and B are 0.25 multiplied by 1,200 kg, or 300 kg, and 0.4 multiplied by 8,000 kg, or 3,200 kg, respectively.

The final step (fig. 9, step 4) is to sum the weighted attribute values determined in the previous step to calculate the quantitative estimate of the attribute (atrazine use) for the target zone (Pudding River basin). The 300 kg estimated for county A and 3,200 kg of atrazine use estimated for county B are added, which results in an estimate of 3,500 kg of atrazine used in the Pudding River basin. If area-weighted areal interpolation was used in this example rather than land-cover weighted areal interpolation, the atrazine use estimate would have been 3,120 kg:

$$(0.6 \times 1,200 \text{ kg atrazine}) + (0.3 \times 8,000 \text{ kg atrazine})$$
$$= 720 + 2,400 = 3,120 \text{ kg,}$$

or about 11 percent lower than the (3,500 kg) estimate derived from using land-cover weighted interpolation.

For the most part, applying land-cover-weighted areal interpolation over area-weighted areal interpolation improves estimates of total quantities, but in some cases, the results are identical. In the example above, if the land cover for the two counties, Clackamas and Marion, was entirely cropland, the area-weighted and land-cover-weighted weighting factors would have been the same, thereby producing the same atrazine-use estimates. The same results would be observed if the cropland in both counties comprised multiple polygons, but all fell entirely within the basin. Therefore, the spatial distribution of the selected land classification(s) within the (partially intersecting) source zone in relation to the target zone determines the degree of potential accuracy improvement with the addition of mapped land cover.

Note that if the intention was to acquire atrazine-use intensity over a drainage basin, there are multiple options. One approach would be to divide the pesticide use estimate derived from land-cover-weighted areal interpolation (as illustrated in fig. 9) by the area of the drainage basin. Two alternate approaches would require the preexistence of a pesticide-use-on-cropland raster that was developed by weighting the (county-based) pesticide use on crops over areas of mapped agricultural land (Nakagaki, 2007a,b). These

pesticide use rasters could be used with simple areal overlay to acquire pesticide-use intensity by (1) obtaining the sum of the pesticide-use grid cell values within the basin, then dividing the sum by the basin area, or by (2) obtaining the mean of the pesticide-use grid cell values within the basin. One consideration for deciding whether simple areal overlay would be sufficient is to determine if the resolution of the pesticide-use raster is suitable for the size of the study areas. The size of the study areas alone however, is not a completely reliable indicator for determining the appropriate grid cell resolution (Nakagaki and Wolock, 2005). Although outliers are not common, exceptions do occur. The most suitable method is often decided from the level of precision desired and an acceptable margin of error for the particular application.

Limitations and Related Adjustments

The characteristics compiled for NAWQA sites and associated study areas are subject to a number of limitations that can be grouped as either data-driven or method-driven. Factors that contribute to potential error from the input data include the accuracy of the sampling site locations, study area boundaries, and the inherent quality of the GIS thematic dataset. In regard to the methods, area-weighted interpolation gives rise to potential error based on the assumption of uniform distribution or density of the attribute; additional error is introduced with land-cover weighted interpolation if the mapped land cover and data table of the attributes by source zones have temporal or other differences that stem from the integration of multiple data sources.

Clearly, the likelihood for erroneous site characteristics is high with the use of inaccurate geographic coordinates for the sampling sites. Such errors generally are more pronounced with discrete data than continuous data. This is due to the nature of continuous data compared to the nature of discrete data: the neighboring cells of continuous data are likely to have either the same, or a slightly higher or lower quantitative value than the grid cell intersecting the station, whereas the neighboring cells of discrete data have a higher likelihood of having a different classification. The difference in value of continuous data typically has much less effect on data analyses. The probability for error is also reduced for sampling stations intersected with either discrete or continuous data with larger grid cells. An incorrect site location, however, can jeopardize not only site characteristics but also well buffer and drainage basin characteristics because these boundaries are developed from the location of the sampling site.

Similarly, inaccurate study-area characteristics are likely to result from the use of inaccurate study-area boundaries. There are different degrees of inaccuracy; for example, a drainage basin boundary derived from topographic and hydrological data at 1:100,000-scale resolution, rather than at the 1:24,000-scale, is insignificant compared to a drainage basin boundary that unintentionally included a noncontributing sub-basin that only contributes to the drainage area during major storms. The latter would have a greater effect on basin characteristics because, for instance, the difference in area could alter the percentages of land cover composition enough to alter the basin's major land classification, whereas improving boundaries with higher resolution source data might change only a fraction of the land-cover percentage for the basin.

Additionally, inaccuracies in the GIS thematic dataset will adversely affect the reliability of site and study-area characteristics. Even though the NAWQA program utilized what was believed to be the best quality and most consistent national scale data available, all thematic datasets, in general, have uncertainty associated with them. Furthermore, in many cases the extent of uncertainty is unknown. Whether or not errors introduced are significant depends on the application. For example, if a researcher wishes to know only if land cover in a basin is "green vegetation" or "not green vegetation," then a miscoding in the land cover dataset of "deciduous forest" instead of "orchards/vineyards" is not a problem. However, if the researcher wants an estimate of pesticides applied to orchards and vineyards in the basin, then a miscoding of "deciduous forest" instead of "orchards/vineyards" would be a concern because the error could result in misrepresented agricultural pesticide use for the study area.

Potential errors of study-area characteristics can also be introduced from the methods, such as the assumption of homogeneous distribution of an attribute across source zones with area-weighted interpolation. (Note: this limitation pertains only to partially intersecting source zones within a target zone and not to the source zones entirely within the target zone.) In other words, the reliability of target zone estimates acquired from area-weighted areal interpolation depends on the degree of heterogeneity across the source zones that partially intersect the target zone: the lesser degree of heterogeneity, the lesser likelihood of error.

Errors associated with area-weighted areal interpolation can be minimized by applying the smallest units of source zones possible, such as census blocks in place of counties or census block groups. The census block is the smallest geographic entity for which the U.S. Census Bureau collects; in 1990, there were approximately 7 million census blocks in the conterminous United States compared to about 230,600 census block groups (Hitt, 1994). The constraints to using smaller source zones, however, include the absence of, or difficulty in attaining, readily available mapped and refined source zones, as well as computer-related issues, such as GIS software limitations, additional costs to acquire high performance processors, and large amounts of data storage.

Although land-cover-weighted areal interpolation is a vast improvement over area-weighted interpolation in general, a typical limitation is disagreement in the area of the land categories from mapped land cover (such as area of cropland) compared to tabulated statistics (such as area of cropland on which a pesticide is applied). Unsurprisingly, disagreement arises from the use of multiple data sources because data-collection methods differ and criteria for classifying land differ. The problem is heightened for areas where the mapped land cover is completely missing a land-cover type, whereas the tabulated data indicate otherwise. For example, a drainage basin is located within a county, and the county shows some atrazine use on row crops (say 1,000 acres of cropland), but according to the mapped land cover, there is no cropland in this county. By employing land-cover-weighted interpolation under these conditions, the atrazine use for this county will be completely disregarded and therefore "lost." One approach to avoiding the loss of data (amount of atrazine use, in this case) is to apply area-weighted areal interpolation for these partially intersecting counties in place of land-cover weighted interpolation by assuming a homogeneous atrazine-use intensity throughout the county. Unless the entire county consists of agricultural land, however, this approach would more than likely misrepresent the true distribution of agricultural land and, subsequently, misrepresent the allocation of pesticide use.

The process of integrating geographic data compiled from different time spans is an ongoing challenge associated with characteristics calculated from land-cover-weighted areal interpolation. Seamless mapped land cover for the conterminous United States is available for less frequent intervals than statistics by source zone; consequently, national land cover data that correlate closest to the date of the source-zone data are typically used. NAWQA has extensively applied land-cover-weighted areal interpolation by integrating tabulated county data and mapped land cover that match best temporally, but do not coincide completely. Examples include, but are not limited to, (1) county-based estimates of annual agricultural atrazine use from 1998 through 2007 (Thelin, 2010) apportioned on mapped agricultural land from the NLCD 2001 (U.S. Geological Survey, 2007; Homer and others, 2007) and from the NLCD 2006 (Fry and others, 2011; U.S. Geological Survey, 2011) ; (2) county-based areas of agricultural management practices from the 1992 National Resource Inventory (U.S. Department of Agriculture, 1995) and the 1997 Natural Resource Inventory (U.S. Department of Agriculture, 2000), compiled at the county level and distributed on mapped agricultural land (Michael E. Wieczorek, U.S. Geological Survey, written commun., 2004) from the National Land Cover Data 1992 (U.S. Geological Survey, 1999; Vogelmann and others, 2001) and, (3) county-based estimates of historical pesticides, such as dieldrin, DDT, and chlordane, applied on crops (Gail P. Thelin, U.S. Geological Survey, written commun., 2004), alllocated on mapped agricultural land from the Land Use and Land Cover (LULC) dataset, which is based on aerial photographs taken in the 1970s to mid-1980s (U.S. Geological Survey, 1990; Price and others, 2007).

The reliability of site and study-area characterization will increase as new data sources emerge and existing data sources improve in quality and become available in more frequent intervals. Examples include (1) the refinement of drainage-basin boundaries as higher-quality landscape and mapped hydrography become available; (2) the development of annual pesticide-use estimates by county for the conterminous U.S. from 1992 through 2009 (Gail P. Thelin, written commun., 2012), which will replace the county 1992 and 1997 pesticide-use estimates (Thelin, 2005a,b) and the county 2002 pesticide-use estimates (Gail P. Thelin, written commun., 2007); (3) the completion of the Cropland Data Layer for the conterminous United States (U.S. Department of Agriculture, 2010a; Johnson and Mueller, 2010), which could be used in replacement for the land classified as "cultivated crops" in the National Land Cover Datasets; and (4) the replacement of the 1:250,000-scale State Soil Geographic (STATSGO) database, which was designed for regional and national analyses (U.S. Department of Agriculture, 1994), with 1:12,000-scale to 1:63,360-scale Soil Survey Geographic (SSURGO) database, which was designed for soil analyses at a much smaller scale, such as counties (U.S. Department of Agriculture, 2010b).

Summary

The physical and anthropogenic characteristics that define the environmental settings of NAWQA's sampling sites and study areas are critical variables that support national water-quality assessment. It is essential that the calculation of these characteristics be performed using methods that result in consistency in data development and geoprocessing. This report described the three major methods for site characterization and provided references and online linkages to the national GIS data sources used by NAWQA.

Acknowledgments

The authors acknowledge and appreciate David M. Wolock, for his technical assistance with the methods described in this report. The authors would also like to express our sincere appreciation to the numerous hydrologists and geographers of the U.S. Geological Survey who spent countless hours preparing the GIS datasets of drainage basin and groundwater study-area boundaries used to characterize the sampled sites and study areas analyzed in the NAWQA program. Finally, a number of national geospatial datasets used for characterization have been developed by our colleagues, namely, Nancy T. Baker, Jo Ann M. Gronberg, Andrew E. LaMotte, Barbara C. Ruddy, Gail P. Thelin, Michael E. Wieczorek, and David M. Wolock.

References Cited

Bell, R.W., and Williamson, A.K., 2006, Data delivery and mapping over the web—National Water-Quality Assessment Data Warehouse: U.S. Geological Survey Fact Sheet, 2006–3101, 6 p.

Fisher, P.F., and Langford, M., 1996, Modeling sensitivity to accuracy in classified imagery—A study of areal interpolation by dasymetric mapping: Professional Geographer, v. 48, no. 3, p. 299–309.

Fry, J., Xian, G., Jin, S., Dewitz, J., Homer, C., Yang, L., Barnes, C., Herold, N., and Wickman, J., 2011, Completion of the 2006 national land cover database for the conterminous United States, Photogrammetric Engineering & Remote Sensing, v. 77, no. 9, p. 858–864.

Gilliom, R.J., Alley, W.M., and Gurtz, M.E., 1995, Design of the National Water-Quality Assessment Program—Occurrence and distribution of water-quality conditions: U.S. Geological Survey Circular 1112, 33 p.

Gilliom, R.J., Hamilton, P.A., and Miller, T.L., 2001, The National Water-Quality Assessment Program—Entering a new decade of investigations: U.S. Geological Survey Fact Sheet 071–01, 6 p., accessed August 21, 2003, at http://water.usgs.gov/pubs/FS/fs-071-01/pdf/fs07101.pdf.

Gilliom, R.J., Barbash, J.E., Crawford, C.G., Hamilton, P.E., Martin, J.D., Nakagaki, Naomi, Nowell, L.H., Scott, J.C., Stackelberg, P.E., Thelin, G.P., and Wolock, D.M., 2006, The Quality of Our Nation's Waters—Pesticides in the Nation's Streams and Groundwater, 1992–2001: U.S. Geological Survey Circular 1291, 172 p., accessed August 29, 2007, at http://ca.water.usgs.gov/pnsp/pubs/circ1291/index.html.

Hitt, K.J., 1994, Refining 1970's land-use data with 1990 population data to indicate new residential development: U.S. Geological Survey Water-Resources Investigations Report, 15 p.

Holt, J.B., Lo, C.P., and Hodler, T.W., 2004, Dasymetric estimation of population density and areal interpolation of census data: Cartography and Geographic Information Science, v. 31, no. 2, p. 103–121.

Homer, C., Dewitz, J., Fry, J., Coan, M., Hossain, N., Larson, C., Herold, N., McKerrow, A., VanDriel, J.N., and Wickman, J., 2007, Completion of the 2001 national land cover database for the conterminous United States: Photogrammetric Engineering and Remote Sensing, v. 73, no. 4, p. 337–341.

Johnson, D.M. and Mueller, R., 2010, The 2009 cropland data layer: Photogrammetric Engineering and Remote Sensing, v. 76, no. 11, p. 1201–1205.

Lanfear, K.J., 1991, 1:2,000,000-scale Digital Line Graph files of streams: U.S. Geological Survey digital vector data, accessed November 15, 2007, at http://water.usgs.gov/lookup/getspatial?stream.

Nakagaki, Naomi, 2007a, Grids of agricultural pesticide use in the conterminous United States, 1992: U.S. Geological Survey digital raster data, accessed August 29, 2007, at http://water.usgs.gov/lookup/getspatial?agpest92grd.

Nakagaki, Naomi, 2007b, Grids of agricultural pesticide use in the conterminous United States, 1997: U.S. Geological Survey digital raster data, accessed August 29, 2007, at http://water.usgs.gov/lookup/getspatial?agpest97grd.

Nakagaki, Naomi, 2010, Drainage basins used for development of the watershed regressions for pesticides (WARP) model: U.S. Geological digital vector data, accessed June 4, 2010, at http://water.usgs.gov/lookup/getspatial?warpbas2010.

Nakagaki, Naomi, and Wolock, D.M., 2005, Estimation of agricultural pesticide use in drainage basins using land cover maps and county pesticide data: U.S. Geological Survey Open-File Report, 2005–1188, 46 p., accessed August 29, 2007, at http://pubs.usgs.gov/of/2005/1188/.

National Atlas of the United States, 2005, 2000 County boundaries of the United States: National Atlas of the United States, Reston, Va., U.S. Geological Survey digital vector data, accessed July 19, 2005, at http://nationalatlas.gov/atlasftp.html.

Nolan, B.T., and Hitt, K.J., 2006, Vulnerability of shallow groundwater and drinking-water wells to nitrate in the United States: Environmental Science and Technology, v. 40, no. 24, p. 7834–7840, accessed December 10, 2007, at http://water.usgs.gov/nawqa/nutrients/pubs/est_v40_no24/.

Nowell, L.H., Crawford, C.G., Nakagaki, N., Thelin, G.P., and Wolock, D.M., 2006, Regression model for explaining and predicting concentrations of dieldrin in whole fish from United States streams: U.S. Geological Survey Scientific Investigations Report 2006–5020, 30 p., accessed August 29, 2007, at http://pubs.usgs.gov/sir/2006/5020.

Price, C.V., Nakagaki, Naomi, Hitt, K.J., and Clawges, R.M., 2007, Enhanced historical land-use and land-cover data sets of the U.S. Geological Survey: U.S. Geological Survey Data Series 240, accessed August 29, 2007, at http://pubs.usgs.gov/ds/2006/240.

Price, C.V., Nakagaki, Naomi, and Hitt, K. J., 2010, National Water-Quality Assessment (NAWQA) Area-Characterization Toolbox, Release 1.0, U.S. Geological Survey Open-File Report 2010–1268 [online-only], accessed November 5, 2012, at http://pubs.usgs.gov/ofr/2010/1268.

Riebel, M., and Agrawal, A., 2007, Areal interpolation of population counts using pre-classified land cover data: Population Research and Policy Review, v. 26, p. 619–33.

Squillace, P.J., and Price, C.V., 1996, Urban land-use study plan for the National Water-Quality Assessment Program: U.S. Geological Survey Open-File Report 96–217, 19 p., accessed November 5, 2012, at http://pubs.usgs.gov/of/1996/0217/report.pdf.

Stackelberg, P.E., Gilliom, R.J., Wolock, D.M., and Hitt, K.J., 2006, Development and application of a regression equation for estimating the occurrence of atrazine in shallow groundwater beneath agricultural areas of the United States: U.S. Geological Survey Scientific Investigations Report 2005–5287, 27 p., accessed August 29, 2007, at http://pubs.usgs.gov/sir/2005/5287.

Stone, W.W., Gilliom, R.J., and Crawford, C.G., 2008, Watershed Regressions for Pesticides (WARP) for Predicting Annual Maximum and Maximum Moving-Average Concentrations of Atrazine in Streams: U.S. Geological Survey Open-File Report 08–1186, 19 p.

Sun, R.J., Weeks, J.B., and Hayes, F.G., 1997, Bibliography of Regional Aquifer-System Analysis Program of the U.S. Geological Survey, 1978–96: U.S. Geological Survey Water-Resources Investigations Report 97–4074, accessed November 29, 2007, at http://water.usgs.gov/ogw/rasa/html/introduction.html.

Thelin, G.P., 2005a, 1992 County pesticide use estimates for 200 compounds (ver. 2.0), U.S. Geological Survey digital file, accessed October 27, 2005, at http://water.usgs.gov/lookup/getspatial?pesticide_use92.

Thelin, G.P., 2005b, 1997 County pesticide use estimates for 220 compounds (ver. 2.0), U.S. Geological Survey digital file, accessed October 27, 2005, at http://water.usgs.gov/lookup/getspatial?pesticide_use97.

Thelin, G.P., 2010, Annual county atrazine use estimates for agriculture (ver. 1.1), digital tabular file, accessed September 22, 2010, at http://water.usgs.gov/lookup/getspatial?sir2010_5034.

U.S. Department of Agriculture, 1994, State soil geographic (STATSGO) database —data use information: Miscellaneous publication no. 1492, (rev. ed.): Fort Worth, Texas, Natural Resources Conservation Service [variously paged].

U.S. Department of Agriculture, 1995, 1992 National Resources Inventory: Natural Resources Conservation Service, Washington, D.C., and Statistical Laboratory, Iowa State University, Ames, Iowa, [CD-ROM].

U.S. Department of Agriculture, 2000 (Reissued 2001), 1997 National Resources Inventory: Natural Resources Conservation Service, Washington, DC, and Statistical Laboratory, Iowa State University, Ames, Iowa, [CD-ROM].

U.S. Department of Agriculture, 2010a, 2009 Cropland data layer: National Agricultural Statistics Service, accessed December 30, 2011, at http://www.nass.usda.gov/research/Cropland/SARS1a.htm.

U.S. Department of Agriculture, 2010b, Soil Survey Geographic (SSURGO) Database: Soil Survey Staff, Natural Resources Conservation Service, accessed October 18, 2010, at http://soildatamart.nrcs.usda.gov.

U.S. Geological Survey, 1990, Land use and land cover digital data from 1:250,000- and 1:100,000-scale maps: U.S. Geological Survey Data User Guide, no. 4, 25 p.

U.S. Geological Survey, 1999, National land cover data 1992, accessed June 16, 2005, at http://www.mrlc.gov/nlcd92_data.php.

U.S. Geological Survey, 2000, Groundwater atlas of the United States: U.S. Geological Survey Hydrologic Atlas 730, accessed November 29, 2007, at http://capp.water.usgs.gov/gwa/gwa.html.

U.S. Geological Survey, 2003, National elevation dataset (NED), accessed April 2003, currently at http://nationalmap.gov/elevation.html.

U.S. Geological Survey, 2007, National land cover database 2001, accessed May 25, 2007, at http://www.mrlc.gov/mrlc2k_nlcd.asp.

U.S. Geological Survey, 2010, USGS National Water Quality Assessment Data Warehouse, accessed February 25, 2010, at http://water.usgs.gov/nawqa/data.

U.S. Geological Survey, 2011, National land cover database 2006 (NLCD 2006), accessed April 2011, at http://www.mrlc.gov/nlcd06_data.php.

Vogelmann, J.E., Howard, S.M., Yang, L., Larson, C.R., Wylie, B.K., and Van Driel, N., 2001, Completion of the 1990's national land cover data set for the conterminous United States from Landsat Environmental Mapper data and ancillary data sources: Photogrammetric Engineering and Remote Sensing, v. 67, p. 650–662.

Appendix A. The National GIS Thematic Datasets Used by the National Water-Quality Assessment (NAWQA) Program to Characterize Sampling Sites for Streams and Groundwater

Appendix A. The national GIS thematic datasets used by the National Water-Quality Assessment (NAWQA) program to characterize sampling sites for streams and groundwater.

[Abbreviations. AVHRR, Advanced Very High Resolution Radiometer; CDL, Cropland Data Layer; CTIC, Conservation Technology Information Center; DEM, Digital Elevation Model; GIS, geographic information system; LULC, Land Use and Land Cover; MSA, Metropolitan Statistical Areas; NADP, National Atmospheric Deposition Program; NAWQA, National Water-Quality Assessment (program); NCDC, National Climatic Data Center; NED, National Elevation Data; NHDPlus, National Hydrography Dataset Plus; NLCD 2001, National Land Cover Database 2001; NLCD 2006, National Land Cover Database 2006; NLCDe 92, Enhanced version of the National Land Cover Data 1992; NLCDeP, Enhanced National Land Cover Data 1992 revised with 1990 and 2000 population data; NPDES, National Pollutant Discharge Elimination System; NRI, National Resource Inventory; PAD, Protected Area Database; PRISM, Parameter-elevation Regressions on Independent Slopes Model; RUSLE, Revised Universal Soil Loss Equation; SSURGO, Soil Survey Geographic (database); STATSGO, State Soil Geographic (database); STATSGO2, U.S. General Soil Map (database); TIGER, Topologically Integrated Geographic Encoding and Referencing (system); TRI, Toxic Release Inventory; USEPA, U.S. Environmental Protection Agency; m, meters; km, kilometers; ~, approximately; –, not applicable]

Method used by NAWQA	General category	Characteristic	Description of the national thematic dataset	Data format, and scale or resolution of dataset used by NAWQA	Original data format, scale or resolution, if different	Data references and miscellaneous notes
Simple overlay	Anthropogenic parameters	Number and density of dams	Location of dams for decadal time periods from 1940 to present	1:100,000-scale points	–	From the National Inventory of Dams (U.S. Army Corps of Engineers, 2006).
Simple overlay	Anthropogenic parameters	Percent impervious surface area	Percent cover of man-made impervious surface	1-km resolution raster	–	From impervious surface areas developed by combining satellite observed nighttime lights, Landsat-derived land cover, and U.S. Census Bureau road vectors (National Oceanic and Atmospheric Administration, 2006).
Simple overlay	Anthropogenic parameters	Density of point sources	Location of discharge sites	1:100,000-scale points	–	From the NPDES program (U.S. Environmental Protection Agency, 2006).
Simple overlay	Anthropogenic parameters	Mean population density in 1970s urban areas	1990 population density by census block group geography in 1970s urban areas	100-m resolution raster	1990 census block group boundaries: 1:100,000-scale polygons; Urban areas: 1:250,000- and 1:100,000-scale polygons	Developed by combining mapped urban areas in the 1970s with 1990 population density from the U.S. Census Bureau. Urban-classified polygons from the 1970s LULC dataset[12] (U.S. Geological Survey, 1990, 1998) were rasterized at the 100-m resolution. The resulting urban-classified grid cells were overlain with the 1990 population density raster compiled by Price (2003).
Simple overlay	Anthropogenic parameters	Mean population density in 1990s urban areas	1990 population density by census block group in 1990s urban areas	Population Density: 100-m resolution raster, 1990s urban areas: 30-m resolution raster	1990 census block group boundaries: 1:100,000-scale polygons	Developed by combining mapped urban areas in the 1990s and 1990 population density from the U.S. Census Bureau. Urban areas extracted from the 30-m resolution NLCDe 92[3,4] (Nakagaki and others, 2007) were resampled at the 100-m resolution. The resulting urban-classified grid cells were overlain with the 1990 population density raster compiled by Price (2003).
Simple overlay	Anthropogenic parameters	Mean population density: 1990	1990 population density by census block group geography	1-km resolution raster	1990 census block group boundaries: 1:100,000-scale polygons; land cover from the 1990 AVHRR satellite data: ~1-km resolution raster	From Price and Clawges (1999a), which is a compilation of 1990 population density by census block groups, and mapped land cover from 1990 AVHRR satellite data (Consortium for International Earth Science Information Network, 1995).
Simple overlay	Anthropogenic parameters	Mean population density: 1990	1990 population density by census block group geography	100-m resolution raster	1990 census block group boundaries: 1:100,000-scale polygons	From Price (2003), which is a gridded representation of mapped 1990 census block groups from 1990 TIGER/Line files and tabular data of population and land areas by block group from the U.S. Census Bureau.

Appendix A. The national GIS thematic datasets used by the National Water-Quality Assessment (NAWQA) program to characterize sampling sites for streams and groundwater.—Continued

[**Abbreviations**: AVHRR, Advanced Very High Resolution Radiometer; CDL, Cropland Data Layer; CTIC, Conservation Technology Information Center; DEM, Digital Elevation Model; GIS, geographic information system; LULC, Land Use and Land Cover; MSA, Metropolitan Statistical Areas; NADP, National Atmospheric Deposition Program; NAWQA, National Water-Quality Assessment (program); NCDC, National Climatic Data Center; NED, National Elevation Data; NHDPlus, National Hydrography Dataset Plus; NLCD 2001, National Land Cover Database 2001; NLCD 2006, National Land Cover Database 2006; NLCDe 92, Enhanced version of the National Land Cover Data 1992; NLCDeP, Enhanced National Land Cover Data 1992 revised with 1990 and 2000 population data; NPDES, National Pollutant Discharge Elimination System; NRI, National Resource Inventory; PAD, Protected Area Database; PRISM, Parameter-elevation Regressions on Independent Slopes Model; RUSLE, Revised Universal Soil Loss Equation; SSURGO, Soil Survey Geographic (database); STATSGO, State Soil Geographic (database); STATSGO2, U.S. General Soil Map (database); TIGER, Topologically Integrated Geographic Encoding and Referencing (system); TRI, Toxic Release Inventory; USEPA, U.S. Environmental Protection Agency: m, meters; km, kilometers; ~, approximately; –, not applicable]

Method used by NAWQA	General category	Characteristic	Description of the national thematic dataset	Data format, and scale or resolution of dataset used by NAWQA	Original data format, scale or resolution, if different	Data references and miscellaneous notes
Area-weighted areal interpolation	Anthropogenic parameters	Mean population density: 1990	1990 population density by census block group geography	1990 census block group boundaries: 30-m resolution raster; 1990 population density: tabular by census block group	1990 census block group boundaries: 1:100,000-scale polygons	Developed by combining a gridded representation of the vector polygons of the 1990 census block group boundaries (U.S. Census Bureau, 2001a) and a data table of 1990 population density by block group compiled by Hitt (1992).
Simple overlay	Anthropogenic parameters	Mean population density: 1990 and 2000	1990 and 2000 population density by census block geography	100-m resolution raster	1990 and 2000 census block boundaries: 1:100,000-scale polygons, 1990 and 2000 population and density: tabular by census block	From U.S. Census Bureau's compilation of population density for 1990 and 2000, compiled by the SILVIS Laboratory (2007) of the Forest and Wildlife Ecology group at the University of Wisconsin-Madison, WI (Radeloff and others, 2005).
Area-weighted areal interpolation	Anthropogenic parameters	Mean population density: 2000	2000 population density by census block group geography	2000 census block group boundaries;30-m resolution raster; 2000 population density: tabular by census block group	2000 census block group boundaries: 1:100,000-scale polygons	Developed by combining a gridded representation of the vector polygons of the 2000 block groups (U.S. Census Bureau, 2001b) and a data table of population density by census block groups obtained from Geolytics (2001a).
Simple overlay	Anthropogenic parameters	Mean population density: 2000	2000 population density by census block group geography	100-m resolution raster	2000 census block group boundaries: 1:100,000-scale polygons	From Hitt (2003), which is a gridded representation of mapped 2000 census block groups from 2000 TIGER/Line files and tabular data of population and land areas by block group from the U.S. Census Bureau.
Area-weighted areal interpolation	Anthropogenic parameters	Mean of housing unit variables: 1990	1990 housing unit variables by census block group geography	1990 census block group boundaries: 30-m resolution raster; 1990 census variables: tabular by census block group	1990 census block group boundaries: 1:100,000-scale polygons	Developed by combining a gridded representation of the vector polygons of the 1990 census block groups (U.S. Census Bureau, 2001a) and a data table of housing unit variables by census block groups (U.S. Bureau of the Census, 1992). Selected variables include sources of water (public system or private company, individual drilled or dug well, or other), sources of sewage disposal (public sewer, septic tank or cesspool, or other), and age of housing (median year built).

Appendix A. The national GIS thematic datasets used by the National Water-Quality Assessment (NAWQA) program to characterize sampling sites for streams and groundwater.—Continued

[Abbreviations: AVHRR, Advanced Very High Resolution Radiometer; CDL, Cropland Data Layer; CTIC, Conservation Technology Information Center; DEM, Digital Elevation Model; GIS, geographic information system; LULC, Land Use and Land Cover; MSA, Metropolitan Statistical Areas; NADP, National Atmospheric Deposition Program; NAWQA, National Water-Quality Assessment (program); NCDC, National Climatic Data Center; NED, National Elevation Data; NHDPlus, National Hydrography Dataset Plus; NLCD 2001, National Land Cover Database 2001; NLCD 2006, National Land Cover Database 2006; NLCDe 92, Enhanced version of the National Land Cover Data 1992; NLCDeP, Enhanced National Land Cover Data 1992 revised with 1990 and 2000 population data; NPDES, National Pollutant Discharge Elimination System; NRI, National Resource Inventory; PAD, Protected Area Database; PRISM, Parameter-elevation Regressions on Independent Slopes Model; RUSLE, Revised Universal Soil Loss Equation; SSURGO, Soil Survey Geographic (database); STATSGO2, U.S. General Soil Map (database); TIGER, Topologically Integrated Geographic Encoding and Referencing (system); TRI, Toxic Release Inventory; USEPA, U.S. Environmental Protection Agency; m, meters; km, kilometers; ~, approximately; —, not applicable]

Method used by NAWQA	General category	Characteristic	Description of the national thematic dataset	Data format, and scale or resolution of dataset used by NAWQA	Original data format, scale or resolution, if different	Data references and miscellaneous notes
Simple overlay	Anthropogenic parameters	Mean housing unit density: 2000	2000 housing unit density by census block geography	100-m resolution raster	2000 census block boundaries: 1:100,000-scale polygons; 2000 housing unit density: tabular by census block	From the U.S. Census Bureau's compilation of housing unit density, compiled by the SILVIS Laboratory (2007) of the Forest and Wildlife Ecology group at the University of Wisconsin-Madison, WI (Radeloff and others, 2005).
Simple overlay	Anthropogenic parameters	Mean population density: 2010	2010 population density by census block group geography	30-m resolution raster	2010 census block group boundaries: 1:100,000-scale; 2010 population density: tabular by census block group	Developed by combining the 2010 TIGER/Line files and tabular data of population and land areas by block group (U.S. Census Bureau, 2011; Curtis V. Price, U.S. Geological Survey, written commun., November 2011).
Simple overlay	Anthropogenic parameters	Metropolitan areas	Metropolitan areas greater than 50 persons per square kilometer	~1:500,000-scale polygons	—	From Price and Clawges (1999b). Based on metropolitan areas greater than 50 persons per square kilometer compiled from contiguous areas of consolidated MSAs and MSAs, and 1990 population density assembled by "CIESIN", the Consortium for International Earth Science Information Network (1995).
Simple overlay	Anthropogenic parameters	Density of road and stream intersections	Roads and streams	1:100,000-scale lines	—	Road and stream intersection density was calculated from the number of intersections of roads and streams in the study area divided by the total length of streams in the study area. Roads are from the TIGER system/Line roads database (U.S. Census Bureau, 2011) obtained from GeoLytics, Inc. (2001b) and streams are from the NHDPlus (U.S. Geological Survey and U.S. Environmental Protection Agency, 2010; U.S. Environmental Protection Agency, 2008). Road and stream intersection density was calculated from the number of intersections of roads and streams in the study area divided by the total length of streams in the study area.
Simple overlay	Anthropogenic parameters	Road density	Roads	1:100,000-scale lines	—	Road density was calculated from the total length of roads in the study area divided by the area of the study area. Roads are from the TIGER/Line roads database (U.S. Census Bureau, 2011) obtained from GeoLytics, Inc. (2001b).

Appendix A. The national GIS thematic datasets used by the National Water-Quality Assessment (NAWQA) program to characterize sampling sites for streams and groundwater.—Continued

[**Abbreviations**: AVHRR, Advanced Very High Resolution Radiometer; CDL, Cropland Data Layer; CTIC, Conservation Technology Information Center; DEM, Digital Elevation Model; GIS, geographic information system; LULC, Land Use and Land Cover; MSA, Metropolitan Statistical Areas; NADP, National Atmospheric Deposition Program; NAWQA, National Water-Quality Assessment (program); NCDC, National Climatic Data Center; NED, National Elevation Data; NHDPlus, National Hydrography Dataset Plus; NLCD 2001, National Land Cover Database 2001; NLCD 2006, National Land Cover Database 2006; NLCDe 92, Enhanced version of the National Land Cover Data 1992; NLCDeP, Enhanced National Land Cover Data 1992 revised with 1990 and 2000 population data; NPDES, National Pollutant Discharge Elimination System; NRI, National Resource Inventory; PAD, Protected Area Database; PRISM, Parameter-elevation Regressions on Independent Slopes Model; RUSLE, Revised Universal Soil Loss Equation; SSURGO, Soil Survey Geographic (database); STATSGO, State Soil Geographic (database); STATSGO2, U.S. General Soil Map (database); TIGER, Topologically Integrated Geographic Encoding and Referencing (system); TRI, Toxic Release Inventory; USEPA, U.S. Environmental Protection Agency; m, meters; km, kilometers; ~, approximately; –, not applicable]

Method used by NAWQA	General category	Characteristic	Description of the national thematic dataset	Data format, and scale or resolution of dataset used by NAWQA	Original data format, scale or resolution, if different	Data references and miscellaneous notes
Simple overlay	Anthropogenic parameters	Number of storage tanks	Locations of storage tanks attributed with counts of underground, leaking underground, above ground, and total storage tanks	Points at unknown scale	–	From the 1998 StarView database (Vista Information Solutions, Inc., 1999).
Simple overlay	Anthropogenic parameters	Toxic release estimates to air, water and land: 1995	Locations of USEPA-regulated facilities attributed with the amount of of volatile organic compounds to air, water, and land, in 1995	Points at unknown scale	Variable	From Price and Clawges (1999c), which is a compilation of selected attributes from the TRI database (U.S. Environmental Protection Agency, 2010a), Envirofacts for 1995.
Land-cover weighted areal interpolation	Chemical applications	Agricultural pesticide use estimates: 1992	Estimates of pesticides used on agricultural land in 1992	Pesticide use: tabular by county; agricultural land: 30-m resolution raster; county boundaries: 30-m resolution raster	County boundaries: 1:100,000-scale polygons and 1:70,000-scale lines	Derived from 1992 county estimates of pesticide use by crop (Thelin, 2005a) combined with mapped 1990 county boundaries[5] and mapped agricultural land from the NLCDe 92[3,6] (Nakagaki and others, 2007).
Simple overlay	Chemical applications	Agricultural pesticide use estimates: 1992	Estimates of pesticides used on agricultural land in 1992	1-km resolution rasters	Pesticide use: tabular by county; agricultural land: 30-m resolution raster; county boundaries: 1:100,000-scale polygons and 1:70,000-scale lines	Derived from rasters of 1992 agricultural pesticide use (Nakagaki, 2007a), developed from the 1992 county estimates of pesticide use by crop (Thelin, 2005a), mapped 1990 county boundaries[5] and mapped agricultural land from the NLCDe 92[3,6] (Nakagaki and others, 2007). Data are available for 199 compounds.
Land-cover weighted areal interpolation	Chemical applications	Agricultural pesticide use estimates: 1997	Estimates of pesticides used on agricultural land in 1997	Pesticide use: tabular by county; agricultural land: 30-m resolution raster; county boundaries: 30-m resolution raster	County boundaries: 1:100,000-scale polygons and 1:70,000-scale lines	Derived from 1997 county estimates of pesticide use by crop (Thelin, 2005b) combined with mapped 1990 county boundaries[5] and mapped agricultural land from the NLCDe 92[3,6] (Nakagaki and others, 2007). Data are available for 219 compounds.

Appendix A. The national GIS thematic datasets used by the National Water-Quality Assessment (NAWQA) program to characterize sampling sites for streams and groundwater.—Continued

[**Abbreviations**: AVHRR, Advanced Very High Resolution Radiometer; CDL, Cropland Data Layer; CTIC, Conservation Technology Information Center; DEM, Digital Elevation Model; GIS, geographic information system; LULC, Land Use and Land Cover; MSA, Metropolitan Statistical Areas; NADP, National Atmospheric Deposition Program; NAWQA, National Water-Quality Assessment (program); NCDC, National Climatic Data Center; NED, National Elevation Data; NHDPlus, National Hydrography Dataset Plus; NLCD 2001, National Land Cover Database 2001; NLCD 2006, National Land Cover Database 2006; NLCDe 92, Enhanced version of the National Land Cover Data 1992; NLCDeP, Enhanced National Land Cover Data 1992 revised with 1990 and 2000 population data; NPDES, National Pollutant Discharge Elimination System; NRI, National Resource Inventory; PAD, Protected Area Database; PRISM, Parameter-elevation Regressions on Independent Slopes Model; RUSLE, Revised Universal Soil Loss Equation; SSURGO, Soil Survey Geographic (database); STATSGO, State Soil Geographic (database); STATSGO2, U.S. General Soil Map (database); TIGER, Topologically Integrated Geographic Encoding and Referencing (system); TRI, Toxic Release Inventory; USEPA, U.S. Environmental Protection Agency; m, meters; km, kilometers; ~, approximately; –, not applicable]

Method used by NAWQA	General category	Characteristic	Description of the national thematic dataset	Data format, and scale or resolution of dataset used by NAWQA	Original data format, scale or resolution, if different	Data references and miscellaneous notes
Simple overlay	Chemical applications	Agricultural pesticide use estimates: 1997	Estimates of pesticides used on agricultural land in 1997	1-km resolution rasters	Pesticide use: tabular by county; Agricultural land: 30-m resolution raster; County boundaries: 1:100,000-scale polygons and 1:70,000-scale lines	Derived from rasters of 1997 agricultural pesticide use (Nakagaki, 2007b), developed from the 1997 county estimates of pesticide use by crop (Thelin, 2005b), mapped 1990 county boundaries[5] and mapped agricultural land from the NLCDe 92[6,6] (Nakagaki and others, 2007).
Land-cover weighted areal interpolation	Chemical applications	Agricultural atrazine use estimates: 1992–2007	Estimates of pesticides used on agricultural land annually from 1992 through 2007	Atrazine use: tabular by county; agricultural land: 30-m resolution rasters; county boundaries: 30-m resolution rasters	County boundaries: 1:100,000- scale polygons and 1:70,000-scale lines	Derived from 1992–2007 county estimates of atrazine use by crop (Thelin, 2010), combined with mapped county boundaries and mapped agricultural land. For atrazine data years 1992–1997, the NLCDe 92[3] (Nakagaki and others, 2007) was used for the source for mapped agricultural land[6], and combined with mapped 1990 county boundaries[5]. for atrazine data years 1998–2007, the NLCD 2001[7] (U.S. Geological Survey, 2007a; Homer and others, 2007; LaMotte, 2008a,b,c,d) was used for the source for mapped agricultural land[8], and combined with mapped 2001 county boundaries[9].
Simple overlay	Chemical applications	Agricultural atrazine use estimates: 1992–2007	Estimates of pesticides used on agricultural land annually from 1992 through 2007	1-km resolution rasters	Atrazine use: tabular by county; Agricultural land: 30-m resolution rasters; County boundaries: 1:100,000-scale polygons and 1:70,000-scale lines	Derived from 1992–2007 county estimates of atrazine use by crop (Thelin, 2010), combined with mapped county boundaries and mapped agricultural land. For atrazine data years 1992–1997, the NLCDe 92[3] (Nakagaki and others, 2007) was used for the source for mapped agricultural land[6], and combined with mapped 1990 county boundaries[5]; for atrazine data years 1998–2007, the NLCD 2001[7] (U.S. Geological Survey, 2007a; Homer and others, 2007; LaMotte, 2008a,b,c,d) was used for the source for mapped agricultural land[8], and combined with mapped 2001 county boundaries[9].

Appendix A. The national GIS thematic datasets used by the National Water-Quality Assessment (NAWQA) program to characterize sampling sites for streams and groundwater.—Continued

[Abbreviations: AVHRR, Advanced Very High Resolution Radiometer; CDL, Cropland Data Layer; CTIC, Conservation Technology Information Center; DEM, Digital Elevation Model; GIS, geographic information system; LULC, Land Use and Land Cover; MSA, Metropolitan Statistical Areas; NADP, National Atmospheric Deposition Program; NAWQA, National Water-Quality Assessment (program); NCDC, National Climatic Data Center; NED, National Elevation Data; NHDPlus, National Hydrography Dataset Plus; NLCD 2001, National Land Cover Database 2001; NLCD 2006, National Land Cover Database 2006; NLCDe 92, Enhanced version of the National Land Cover Data 1992; NLCDeP, Enhanced National Land Cover Data 1992 revised with 1990 and 2000 population data; NPDES, National Pollutant Discharge Elimination System; NRI, National Resource Inventory; PAD, Protected Area Database; PRISM, Parameter-elevation Regressions on Independent Slopes Model; RUSLE, Revised Universal Soil Loss Equation; SSURGO, Soil Survey Geographic (database); STATSGO, State Soil Geographic (database); STATSGO2, U.S. General Soil Map (database); TIGER, Topologically Integrated Geographic Encoding and Referencing (system); TRI, Toxic Release Inventory; USEPA, U.S. Environmental Protection Agency; m, meters; km, kilometers; ~, approximately; –, not applicable]

Method used by NAWQA	General category	Characteristic	Description of the national thematic dataset	Data format, and scale or resolution of dataset used by NAWQA	Original data format, scale or resolution, if different	Data references and miscellaneous notes
Land-cover weighted areal interpolation	Chemical applications	Historical agricultural pesticide use estimates: 1960s	Estimates of DDT, dieldrin, and chlordane used on agricultural land in the 1960s	Organochlorine use: tabular by county; agricultural land: 100-m resolution raster; county boundaries: 100-m resolution raster	Agricultural land: 1:250,000-scale polygons; county boundaries: 1:100,000-scale polygons and 1:70,000-scale lines	Derived from county estimates of historical organochlorine pesticide use on crops (Gail P. Thelin, U.S. Geological Survey, written commun., 2004), combined with mapped 1990 county boundaries[3] and mapped agricultural land from the 1970s LULC dataset[1,10] (U.S. Geological Survey, 1990, 1998).
Simple overlay	Chemical applications	Historical agricultural pesticide use estimates: 1960s	Estimates of DDT, dieldrin, and chlordane used on agricultural land in the 1960s	1-km resolution rasters	Organochlorine use: tabular by county; Agricultural land: 1:250,000-scale polygons; County boundaries: 1:100,000-scale polygons and 1:70,000-scale lines	Derived from county estimates of historical organochlorine pesticide use on crops (Gail P. Thelin, U.S. Geological Survey, written commun., 2004), combined with mapped 1990 county boundaries[3] and mapped agricultural land from the 1970s LULC dataset[1,10] (U.S. Geological Survey, 1990, 1998).
Simple overlay	Chemical applications	Nitrogen inputs from atmospheric deposition: 1985–2004	Estimates of inorganic nitrogen from atmospheric deposition for 1985 through 2004, based on wet deposition of nitrate and ammonium	1-km resolution rasters	Points	Derived from spatially interpolating nitrogen measurements of atmospheric deposition collected at monitoring sites (points) for the NADP, to 1-km resolution rasters. Nitrogen measurements for sites monitored from 1985 to 2001 (National Atmospheric Deposition Program, 2002) were spatially interpolated to 1-km resolution rasters (Barbara C. Ruddy, U.S. Geological Survey, written commun., 2005); nitrogen measurements for sites monitored from 2002 to 2004 (National Atmospheric Deposition Program, 2005) were also spatially interpolated to 1-km resolution rasters (Jo Ann M. Gronberg, U.S. Geological Survey, written commun., 2005).

Appendix A.   The national GIS thematic datasets used by the National Water-Quality Assessment (NAWQA) program to characterize sampling sites for streams and groundwater.—Continued

[Abbreviations: AVHRR, Advanced Very High Resolution Radiometer; CDL, Cropland Data Layer; CTIC, Conservation Technology Information Center; DEM, Digital Elevation Model; GIS, geographic information system; LULC, Land Use and Land Cover; MSA, Metropolitan Statistical Areas; NADP, National Atmospheric Deposition Program; NAWQA, National Water-Quality Assessment (program); NCDC, National Climatic Data Center; NED, National Elevation Data; NHDPlus, National Hydrography Dataset Plus; NLCD 2001, National Land Cover Database 2001; NLCD 2006, National Land Cover Database 2006; NLCDe 92, Enhanced version of the National Land Cover Data 1992; NLCDeP, Enhanced National Land Cover Data 1992 revised with 1990 and 2000 population data; NPDES, National Pollutant Discharge Elimination System; NRI, National Resource Inventory; PAD, Protected Area Database; PRISM, Parameter-elevation Regressions on Independent Slopes Model; RUSLE, Revised Universal Soil Loss Equation; SSURGO, Soil Survey Geographic (database); STATSGO, State Soil Geographic (database); STATSGO2, U.S. General Soil Map (database); TIGER, Topologically Integrated Geographic Encoding and Referencing (system); TRI, Toxic Release Inventory; USEPA, U.S. Environmental Protection Agency; m, meters; km, kilometers; ~, approximately; –, not applicable]

Method used by NAWQA	General category	Characteristic	Description of the national thematic dataset	Data format, and scale or resolution of dataset used by NAWQA	Original data format, scale or resolution, if different	Data references and miscellaneous notes
Land-cover weighted areal inter-polation	Chemical applications	Nitrogen inputs from fertilizer: 1987–2004	Estimates of nitrogen inputs from non-farm and farm fertilizer annually from 1987 through 2004	Nitrogen inputs from fertilizer: tabular by county; County boundaries: 30-m resolution raster; agricultural land: 30-m resolution rasters	County boundaries: 1:100,000-scale polygons and 1:70,000-scale lines	Derived from 1987–2004 county estimates of nitrogen inputs from fertilizer (Jo Ann M. Gronberg and Norman E. Spahr, U.S. Geological Survey, written commun., 2008), which were developed using methodology modified from Ruddy and others (2006). For nitrogen input data years 1987–2001, the NLCDe 92[3] (Nakagaki and others, 2007) was used for the source for mapped urban[11] and agricultural[6] land, and combined with mapped 1990 county boundaries[5]; for nitrogen input data years 2002–2004, the NLCDeP[12] (Hitt, 2008) was used for the source for mapped urban[13] and agricultural[14] land and combined with mapped 2001 county boundaries[9]
Simple overlay	Chemical applications	Nitrogen inputs from fertilizer: 1987–2004	Estimates of nitrogen inputs from non-farm and farm fertilizer annually from 1987 through 2004	1-km resolution rasters	Nitrogen inputs from fertilizer: tabular by county; Agricultural and urban land: 30-m resolution rasters; County boundaries: 1:100,000-scale polygons and 1:70,000-scale lines	Derived from 1987–2004 county estimates of nitrogen inputs from fertilizer (Jo Ann M. Gronberg and Norman E. Spahr, U.S. Geological Survey, written commun., 2008), which were developed using methodology modified from Ruddy and others (2006). For nitrogen input data years 1987–2001, the NLCDe 92[3] (Nakagaki and others, 2007) was used for the source for mapped urban[11] and agricultural[6] land, and combined with mapped 1990 county boundaries[5]; for nitrogen input data years 2002–2004, the NLCDeP[12] (Hitt, 2008) was used for the source for mapped urban[13] and agricultural[14] land and combined with mapped 2001 county boundaries[9]

Appendix A.   The national GIS thematic datasets used by the National Water-Quality Assessment (NAWQA) program to characterize sampling sites for streams and groundwater.—Continued

[**Abbreviations**: AVHRR, Advanced Very High Resolution Radiometer; CDL, Cropland Data Layer; CTIC, Conservation Technology Information Center; DEM, Digital Elevation Model; GIS, geographic information system; LULC, Land Use and Land Cover; MSA, Metropolitan Statistical Areas; NADP, National Atmospheric Deposition Program; NAWQA, National Water-Quality Assessment (program); NCDC, National Climatic Data Center; NED, National Elevation Data; NHDPlus, National Hydrography Dataset Plus; NLCD 2001, National Land Cover Database 2001; NLCD 2006, National Land Cover Database 2006; NLCDe 92, Enhanced version of the National Land Cover Data 1992; NLCDeP, Enhanced National Land Cover Data 1992 revised with 1990 and 2000 population data; NPDES, National Pollutant Discharge Elimination System; NRI, National Resource Inventory; PAD, Protected Area Database; PRISM, Parameter-elevation Regressions on Independent Slopes Model; RUSLE, Revised Universal Soil Loss Equation; SSURGO, Soil Survey Geographic (database); STATSGO, State Soil Geographic (database); STATSGO2, U.S. General Soil Map (database); TIGER, Topologically Integrated Geographic Encoding and Referencing (system); TRI, Toxic Release Inventory; USEPA, U.S. Environmental Protection Agency; m, meters; km, kilometers; ~, approximately; –, not applicable]

Method used by NAWQA	General category	Characteristic	Description of the national thematic dataset	Data format, and scale or resolution of dataset used by NAWQA	Original data format, scale or resolution, if different	Data references and miscellaneous notes
Land-cover weighted areal interpolation	Chemical applications	Nitrogen inputs from manure: 1982, 1987, 1992, 1997, and 2002	Estimates of nitrogen inputs from manure from confined and non-confined farm animals for 1982, 1987, 1992, 1997, and 2002	Nitrogen inputs from manure: tabular by county; county boundaries: 30-m resolution rasters; agricultural land: 30-m resolution rasters	County boundaries: 1:100,000-scale polygons and 1:70,000-scale lines	Derived from county estimates of nitrogen input from manure based on number of confined and nonconfined farm animals from the Censuses of Agriculture. The 1982, 1987, 1992, and 1997 county data were developed by Ruddy and others (2006) and 2002 county data were developed by David K. Mueller (U.S. Geological Survey, written commun., 2006). County manure data for all years were compiled using the methods described by Ruddy and others (2006). The tabular manure data were integrated with mapped county boundaries and mapped agricultural land. For nitrogen input data years for 1982, 1987, 1992 and 1997, the NLCDe 92[2] (Nakagaki and others, 2007) was used for the source for mapped agricultural land[15] and combined with mapped 1990 county boundaries[5]; for nitrogen input data year 2002, the NLCDeP[12] (Hitt, 2008) was used for the source for mapped agricultural land[16] and combined with mapped 2001 county boundaries[9].
Simple overlay	Chemical applications	Nitrogen inputs from manure: 1982, 1987, 1992, 1997, and 2002	Estimates of nitrogen inputs from manure from confined and non-confined farm animals for 1982, 1987, 1992, 1997, and 2002	1-km resolution rasters	Nitrogen from manure: tabular by county; Agricultural land: 30-m resolution rasters; County boundaries: 1:100,000-scale polygons and 1:70,000-scale lines	Derived from county estimates of nitrogen input from manure based on number of confined and nonconfined farm animals from the Censuses of Agriculture. The 1982, 1987, 1992, and 1997 county data were developed by Ruddy and others (2006) and 2002 county data were developed by David K. Mueller (U.S. Geological Survey, written commun., 2006). County manure data for all years were compiled using the methods described by Ruddy and others (2006). The tabular manure data were integrated with mapped county boundaries and mapped agricultural land. For nitrogen input data years for 1982, 1987, 1992 and 1997, the NLCDe 92[2] (Nakagaki and others, 2007) was used for the source for mapped agricultural land[15] and combined with mapped 1990 county boundaries[5]; for nitrogen input data year 2002, the NLCDeP[12] (Hitt, 2008) was used for the source for mapped agricultural land[16] and combined with mapped 2001 county boundaries[9].

Appendix A.   The national GIS thematic datasets used by the National Water-Quality Assessment (NAWQA) program to characterize sampling sites for streams and groundwater.—Continued

[Abbreviations: AVHRR, Advanced Very High Resolution Radiometer; CDL, Cropland Data Layer; CTIC, Conservation Technology Information Center; DEM, Digital Elevation Model; GIS, geographic information system; LULC, Land Use and Land Cover; MSA, Metropolitan Statistical Areas; NADP, National Atmospheric Deposition Program; NAWQA, National Water-Quality Assessment (program); NCDC, National Climatic Data Center; NED, National Elevation Data; NHDPlus, National Hydrography Dataset Plus; NLCD 2001, National Land Cover Database 2001; NLCD 2006, National Land Cover Database 2006; NLCDe 92, Enhanced version of the National Land Cover Data 1992; NLCDeP, Enhanced National Land Cover Data 1992 revised with 1990 and 2000 population data; NPDES, National Pollutant Discharge Elimination System; NRI, National Resource Inventory; PAD, Protected Area Database; PRISM, Parameter-elevation Regressions on Independent Slopes Model; RUSLE, Revised Universal Soil Loss Equation; SSURGO, Soil Survey Geographic (database); STATSGO, State Soil Geographic (database); STATSGO2, U.S. General Soil Map (database); TIGER, Topologically Integrated Geographic Encoding and Referencing (system); TRI, Toxic Release Inventory; USEPA, U.S. Environmental Protection Agency; m, meters; km, kilometers; ~, approximately; –, not applicable]

Method used by NAWQA	General category	Characteristic	Description of the national thematic dataset	Data format, and scale or resolution of dataset used by NAWQA	Original data format, scale or resolution, if different	Data references and miscellaneous notes
Land-cover weighted areal interpolation	Chemical applications	Phosphorus inputs from fertilizer: 1987–2004	Estimates of phosphorus inputs from non-farm and farm fertilizer annually from 1987 through 2004	Phosphorus inputs from fertilizer: tabular by county; county boundaries: 30-m resolution raster; agricultural land: 30-m resolution rasters	County boundaries: 1:100,000-scale polygons and 1:70,000-scale lines	Derived from 1987–2004 county estimates of phosphorus inputs from fertilizer (Jo Ann M. Gronberg and Norman E. Spahr, U.S. Geological Survey, written commun., 2008), which were developed from methodology modified from Ruddy and others (2006); for phosphorus input data years 1987–2001, the NLCDe 92[3] (Nakagaki and others, 2007) was used for the source for mapped urban[11] and agricultural[6] land and combined with mapped 1990 county boundaries[5]; for phosphorus input data years 2002–2004, the NLCDeP[12] (Hitt, 2008) was used for the source for mapped urban[13] and agricultural[14] land and combined with mapped 2001 county boundaries[9].
Simple overlay	Chemical applications	Phosphorus inputs from fertilizer: 1987–2004	Estimates of phosphorus inputs from non-farm and farm fertilizer annually from 1987 through 2004	1-km resolution rasters	Phosphorus inputs from fertilizer: tabular by county; Agricultural and urban land: 30-m resolution raster; County boundaries: 1:100,000-scale polygons and 1:70,000-scale lines	Derived from 1987–2004 county estimates of phosphorus inputs from fertilizer (Jo Ann M. Gronberg and Norman E. Spahr, U.S. Geological Survey, written commun., 2008), which were developed using methodology modified from Ruddy and others (2006); for phosphorus input data years 1987–2001, the NLCDe 92[3] (Nakagaki and others, 2007) was used for the source for mapped urban[11] and agricultural[6] land and combined with mapped 1990 county boundaries[5]; for phosphorus input data years 2002–2004, the NLCDeP[12] (Hitt, 2008) was used for the source for mapped urban[13] and agricultural[14] land and combined with mapped 2001 county boundaries[9].

Appendix A. The national GIS thematic datasets used by the National Water-Quality Assessment (NAWQA) program to characterize sampling sites for streams and groundwater.—Continued

[**Abbreviations**: AVHRR, Advanced Very High Resolution Radiometer; CDL, Cropland Data Layer; CTIC, Conservation Technology Information Center; DEM, Digital Elevation Model; GIS, geographic information system; LULC, Land Use and Land Cover; MSA, Metropolitan Statistical Areas; NADP, National Atmospheric Deposition Program; NAWQA, National Water-Quality Assessment (program); NCDC, National Climatic Data Center; NED, National Elevation Data; NHDPlus, National Hydrography Dataset Plus; NLCD 2001, National Land Cover Database 2001; NLCD 2006, National Land Cover Database 2006; NLCDe 92, Enhanced version of the National Land Cover Data 1992; NLCDeP, Enhanced National Land Cover Data 1992 revised with 1990 and 2000 population data; NPDES, National Pollutant Discharge Elimination System; NRI, National Resource Inventory; PAD, Protected Area Database; PRISM, Parameter-elevation Regressions on Independent Slopes Model; RUSLE, Revised Universal Soil Loss Equation; SSURGO, Soil Survey Geographic (database); STATSGO, State Soil Geographic (database); STATSGO2, U.S. General Soil Map (database); TIGER, Topologically Integrated Geographic Encoding and Referencing (system); TRI, Toxic Release Inventory; USEPA, U.S. Environmental Protection Agency; m, meters; km, kilometers; ~, approximately; –, not applicable]

Method used by NAWQA	General category	Characteristic	Description of the national thematic dataset	Data format, and scale or resolution of dataset used by NAWQA	Original data format, scale or resolution, if different	Data references and miscellaneous notes
Land-cover weighted areal interpolation	Chemical applications	Phosphorus inputs from manure: 1982, 1987, 1992, 1997, and 2002	Estimates of phosphorus inputs from manure from confined and non-confined farm animals for 1982, 1987, 1992, 1997, and 2002	Phosphorus inputs from manure: tabular by county; county boundaries: 30-m resolution raster; agricultural land: 30-m resolution rasters	County boundaries: 1:100,000-scale polygons and 1:70,000-scale lines	Derived from county estimates of phosphorus input from manure based on number of confined and nonconfined farm animals from the Censuses of Agriculture. The 1982, 1987, 1992, and 1997 county data were developed by Ruddy and others (2006) and 2002 county data were developed by David K. Mueller (U.S. Geological Survey, written commun., 2006). County manure data for all years were compiled using the methods described by Ruddy and others (2006). The tabular manure data were integrated with mapped county boundaries and mapped agricultural land. For phosphorus input data years for 1982, 1987, 1992 and 1997, the NLCDe 92[3] (Nakagaki and others, 2007) was used for the source for mapped agricultural land[15] and combined with mapped 1990 county boundaries[5]; for phosphorus input data year 2002, the NLCDeP[12] (Hitt, 2008) was used for the source for mapped agricultural land[16] and combined with mapped 2001 county boundaries[9].
Simple overlay	Chemical applications	Phosphorus inputs from manure: 1982, 1987, 1992, 1997, and 2002	Estimates of phosphorus inputs from manure from confined and non-confined farm animals for 1982, 1987, 1992, 1997, and 2002	1-km resolution rasters	Phosphorus from manure: tabular by county; Agricultural land: 30-m resolution rasters; County boundaries: 1:100,000-scale polygons and 1:70,000-scale lines	Derived from county estimates of phosphorus input from manure based on number of confined and nonconfined farm animals from the Censuses of Agriculture. The 1982, 1987, 1992, and 1997 county data were developed by Ruddy and others (2006) and 2002 county data were developed by David K. Mueller (U.S. Geological Survey, written commun., 2006). County manure data for all years were compiled using the methods described by Ruddy and others (2006). The tabular manure data were integrated with mapped county boundaries and mapped agricultural land. For phosphorus input data years for 1982, 1987, 1992 and 1997, the NLCDe 92[3] (Nakagaki and others, 2007) was used for the source for mapped agricultural land[15] and combined with mapped 1990 county boundaries[5]; for phosphorus input data year 2002, the NLCDeP[12] (Hitt, 2008) was used for the source for mapped agricultural land[16] and combined with mapped 2001 county boundaries[9].

Appendix A. The national GIS thematic datasets used by the National Water-Quality Assessment (NAWQA) program to characterize sampling sites for streams and groundwater.—Continued

[Abbreviations: AVHRR, Advanced Very High Resolution Radiometer; CDL, Cropland Data Layer; CTIC, Conservation Technology Information Center; DEM. Digital Elevation Model; GIS, geographic information system; LULC, Land Use and Land Cover; MSA, Metropolitan Statistical Areas; NADP, National Atmospheric Deposition Program; NAWQA, National Water-Quality Assessment (program); NCDC, National Climatic Data Center; NED, National Elevation Data; NHDPlus, National Hydrography Dataset Plus; NLCD 2001, National Land Cover Database 2001; NLCD 2006, National Land Cover Database 2006; NLCDe 92, Enhanced version of the National Land Cover Data 1992; NLCDeP, Enhanced National Land Cover Data 1992 revised with 1990 and 2000 population data; NPDES, National Pollutant Discharge Elimination System; NRI, National Resource Inventory; PAD, Protected Area Database; PRISM, Parameter-elevation Regressions on Independent Slopes Model; RUSLE, Revised Universal Soil Loss Equation; SSURGO, Soil Survey Geographic (database); STATSGO, State Soil Geographic (database); STATSGO2, U.S. General Soil Map (database); TIGER, Topologically Integrated Geographic Encoding and Referencing (system); TRI, Toxic Release Inventory; USEPA, U.S. Environmental Protection Agency; m, meters; km, kilometers; ~, approximately; –, not applicable]

Method used by NAWQA	General category	Characteristic	Description of the national thematic dataset	Data format, and scale or resolution of dataset used by NAWQA	Original data format, scale or resolution, if different	Data references and miscellaneous notes
Simple overlay	Climate	Mean of average annual number of consecutive dry days: 1961–1990	Average annual number of consecutive dry days based on the NCDC meteorological stations, defined as the average number of consecutive days there is no rainfall in a single calendar year for the period 1961–1990	1-km resolution raster	Points	Derived from vector points of the NCDC precipitation station data (Williams and others, 2006) used to calculate average annual number of consecutive dry days, then spatially interpolated to a 1-km resolution raster (David M. Wolock, U.S. Geological Survey, written commun., 2001).
Simple overlay	Climate	Mean of average annual number of consecutive wet days: 1961–1990	Average annual number of consecutive wet days based on the NCDC meteorological stations, defined as the average number of consecutive days there is rainfall in a single calendar year for the period 1961–1990	1-km resolution raster	Points	Derived from vector points of the NCDC precipitation station data (Williams and others, 2006) used to calculate average annual number of consecutive wet days, then spatially interpolated to a 1-km resolution raster (David M. Wolock, U.S. Geological Survey, written commun., 2001).
Simple overlay	Climate	Mean of average annual precipitation intensity: 1961–1990	Average annual precipitation intensity for the period 1961–1990 based on the NCDC meteorological stations	1-km resolution raster	Points	Developed from vector points of the NCDC precipitation station data (Williams and others, 2006) used to calculate average annual precipitation intensity, then spatially interpolated to a 1-km resolution raster (David M. Wolock, U.S. Geological Survey, written commun., 2001).
Simple overlay	Climate	Mean of average annual precipitation: 1961–1990	Average annual precipitation for the period 1961–1990, compiled by the PRISM Climate Group	1-km resolution raster	30-arc seconds (~800-m) resolution raster	From the PRISM Climate Group (2010), developed using methods by Daly and others (2002). PRISM's original raster was spatially interpolated to a 1-km resolution raster (David M Wolock, U.S. Geological Survey, written commun., 2001).
Simple overlay	Climate	Mean of average annual temperature: 1961–1990	Average annual temperature for the period 1961–1990, compiled by the PRISM Climate Group	1-km resolution raster	30-arc seconds (~800-m) resolution raster	From the PRISM Climate Group (2010), developed using methods by Daly and others (2002). PRISM's original raster was spatially interpolated to a 1-km resolution raster (David M Wolock, U.S. Geological Survey, written commun., 2001).

Appendix A.   The national GIS thematic datasets used by the National Water-Quality Assessment (NAWQA) program to characterize sampling sites for streams and groundwater.—Continued

[**Abbreviations:** AVHRR, Advanced Very High Resolution Radiometer; CDL, Cropland Data Layer; CTIC, Conservation Technology Information Center; DEM, Digital Elevation Model; GIS, geographic information system; LULC, Land Use and Land Cover; MSA, Metropolitan Statistical Areas; NADP, National Atmospheric Deposition Program; NAWQA, National Water-Quality Assessment (program); NCDC, National Climatic Data Center; NED, National Elevation Data; NHDPlus, National Hydrography Dataset Plus; NLCD 2001, National Land Cover Database 2001; NLCD 2006, National Land Cover Database 2006; NLCDe 92, Enhanced version of the National Land Cover Data 1992; NLCDeP, Enhanced National Land Cover Data 1992 revised with 1990 and 2000 population data; NPDES, National Pollutant Discharge Elimination System; NRI, National Resource Inventory; PAD, Protected Area Database; PRISM, Parameter-elevation Regressions on Independent Slopes Model; RUSLE, Revised Universal Soil Loss Equation; SSURGO, Soil Survey Geographic (database); STATSGO, State Soil Geographic (database); STATSGO2, U.S. General Soil Map (database); TIGER, Topologically Integrated Geographic Encoding and Referencing (system); TRI, Toxic Release Inventory; USEPA, U.S. Environmental Protection Agency; m, meters; km, kilometers; ~, approximately; –, not applicable]

Method used by NAWQA	General category	Characteristic	Description of the national thematic dataset	Data format, and scale or resolution of dataset used by NAWQA	Original data format, scale or resolution, if different	Data references and miscellaneous notes
Simple overlay	Climate	Mean of average monthly precipitation: 1961–1990	Average monthly precipitation for each month for the period 1961–1990, compiled by the PRISM Climate Group	1-km resolution rasters	30-arc seconds (~800-m) resolution raster	From the PRISM Climate Group (2010), developed using methods by Daly and others (2002). PRISM's original raster was spatially interpolated to a 1-km resolution raster (David M Wolock, U.S. Geological Survey, written commun., 2001). A single raster was generated for each month for the period 1961–1990 (total of 12 rasters).
Simple overlay	Climate	Mean of average annual precipitation: 1961–1990	Average annual precipitation for the period 1961–1990, from the NCDC database for 387 climate divisions	1-km resolution raster	Climate division boundaries: ~1:20,000,000-scale polygons; precipitation values: tabular by climate division	Developed from measured precipitation (National Climatic Data Center, 2007) linked to polygons of mapped climate divisions (National Climatic Data Center, 1991) that were rasterized at the 1-km resolution.
Simple overlay	Climate	Mean of average monthly precipitation: 1961–1990	Average monthly precipitation for the period 1961–1990, from the NCDC database for 387 climate divisions	1-km resolution rasters	Climate division boundaries: ~1:20,000,000-scale polygons; precipitation values: tabular by climate division	Developed from measured precipitation (National Climatic Data Center, 2007) linked to polygons of mapped climate divisions (National Climatic Data Center, 1991) that were rasterized at the 1-km resolution. A single raster was generated for each month for the years 1961–1990 (total of 12 rasters).
Simple overlay	Climate	Mean of average precipitation for each month, for years 1961–2009	Average precipitation for each month annually for the period 1961–2009, from the NCDC database for 387 climate divisions	1-km resolution rasters	Climate division boundaries: ~1:20,000,000-scale polygons; precipitation values: tabular by climate division	Developed from measured precipitation (National Climatic Data Center, 2007) linked to polygons of mapped climate divisions (National Climatic Data Center, 1991) that were rasterized at the 1-km resolution. A single raster was generated for each month for the years 1961–2009 (one for each month for the 49 year span, for a total of 588 rasters).
Simple overlay	Climate	Mean of average annual temperature: 1961–1990	Average annual temperature for the period 1961–1990, from the NCDC database for 387 climate divisions	1-km resolution raster	Climate division boundaries: ~1:20,000,000-scale polygons; temperature values: tabular by climate division	Developed from measured temperature (National Climatic Data Center, 2007) linked to polygons of mapped climate divisions (National Climatic Data Center, 1991) that were rasterized at the 1-km resolution.
Simple overlay	Climate	Mean of average monthly temperature: 1961–1990	Average monthly temperature for the period 1961–1990, from the NCDC database for 387 climate divisions	1-km resolution rasters	Climate division boundaries: ~1:20,000,000-scale polygons; temperature values: tabular by climate division	Developed from measured temperature (National Climatic Data Center, 2007) linked to polygons of mapped climate divisions (National Climatic Data Center, 1991) that were rasterized at the 1-km resolution. A single raster was generated for each month for the years 1961–1990 (total of 12 rasters).

Appendix A. The national GIS thematic datasets used by the National Water-Quality Assessment (NAWQA) program to characterize sampling sites for streams and groundwater.—Continued

[**Abbreviations**: AVHRR, Advanced Very High Resolution Radiometer; CDL, Cropland Data Layer; CTIC, Conservation Technology Information Center; DEM, Digital Elevation Model; GIS, geographic information system; LULC, Land Use and Land Cover; MSA, Metropolitan Statistical Areas; NADP, National Atmospheric Deposition Program; NAWQA, National Water-Quality Assessment (program); NCDC, National Climatic Data Center; NED, National Elevation Data; NHDPlus, National Hydrography Dataset Plus; NLCD 2001, National Land Cover Database 2001; NLCD 2006, National Land Cover Database 2006; NLCDe 92, Enhanced version of the National Land Cover Data 1992; NLCDeP, Enhanced National Land Cover Data 1992 revised with 1990 and 2000 population data; NPDES, National Pollutant Discharge Elimination System; NRI, National Resource Inventory; PAD, Protected Area Database; PRISM, Parameter-elevation Regressions on Independent Slopes Model; RUSLE, Revised Universal Soil Loss Equation; SSURGO, Soil Survey Geographic (database); STATSGO, State Soil Geographic (database); STATSGO2, U.S. General Soil Map (database); TIGER, Topologically Integrated Geographic Encoding and Referencing (system); TRI, Toxic Release Inventory; USEPA, U.S. Environmental Protection Agency; m, meters; km, kilometers; ~, approximately; –, not applicable]

Method used by NAWQA	General category	Characteristic	Description of the national thematic dataset	Data format, and scale or resolution of dataset used by NAWQA	Original data format, scale or resolution, if different	Data references and miscellaneous notes
Simple overlay	Climate	Mean of average temperature for each month, for years 1961–2009	Average temperature for each month annually for the period 1961–2009, from the NCDC database for 387 climate divisions	1-km resolution rasters	Climate division boundaries: ~1:20,000,000-scale polygons; temperature values: tabular by climate division	Developed from measured temperature (National Climatic Data Center, 2007) linked to polygons of mapped climate divisions (National Climatic Data Center, 1991) that were rasterized at the 1-km resolution. A single raster was generated for each month for the years 1961–2009 (one for each month for the 49-year span, for a total of 588 rasters).
Simple overlay	Climate	Mean of average annual precipitation: 1961–1990	Average annual precipitation for the period 1961–1990, based on climate normals from the NCDC meteological stations at which daily precipitation time series data have been collected	1-km resolution raster	Points	Developed by spatially interpolating vector points of the 1961–1990 average annual precipitation (Owensby and Ezell, 1992) to a 1-km resolution raster (David M. Wolock, U.S. Geological Survey, written commun., 2001).
Simple overlay	Climate	Mean of average annual temperature: 1961–1990	Average annual temperature for the period 1961–1990, based on climate normals from the NCDC meteorological stations at which daily temperature time series data have been collected	1-km resolution raster	Points	Developed by spatially interpolating vector points of the 1961–1990 climate normals (Owensby and Ezell, 1992) to a 1-km resolution raster (David M. Wolock, U.S. Geological Survey, written commun., 2001).
Simple overlay	Climate	Mean relative humidity: 1961–1990	Relative humidity, derived from 30-years of record, 1961–1990, compiled by the PRISM Climate Group	2-km resolution raster	4-km resolution raster	From the PRISM Climate Group (2010), developed using methods by Daly and others (2002). The original 4-km resolution raster of relative humidity was resampled to 2-km resolution by Climate Source, Inc. (2006; Ryan Hill, Utah State University, personal commun., November 2006).
Simple overlay	Climate	Mean of annual number of days of measurable precipitation: 1961–1990	Annual number of days of measurable precipitation, derived from 30-years of record, 1961–1990, compiled by the PRISM Climate Group	2-km resolution raster	4-km resolution raster	From the PRISM Climate Group (2010), developed using methods by Daly and others (2002). The original 4-km resolution raster was resampled to 2-km resolution by Climate Source, Inc. (2006; Ryan Hill, Utah State University, personal commun., November 2006).

Appendix A.   The national GIS thematic datasets used by the National Water-Quality Assessment (NAWQA) program to characterize sampling sites for streams and groundwater.—Continued

[Abbreviations: AVHRR, Advanced Very High Resolution Radiometer; CDL, Cropland Data Layer; CTIC, Conservation Technology Information Center; DEM, Digital Elevation Model; GIS, geographic information system; LULC, Land Use and Land Cover; MSA, Metropolitan Statistical Areas; NADP, National Atmospheric Deposition Program; NAWQA, National Water-Quality Assessment (program); NCDC, National Climatic Data Center; NED, National Elevation Data; NHDPlus, National Hydrography Dataset Plus; NLCD 2001, National Land Cover Database 2001; NLCD 2006, National Land Cover Database 2006; NLCDe 92, Enhanced version of the National Land Cover Data 1992; NLCDeP, Enhanced National Land Cover Data 1992 revised with 1990 and 2000 population data; NPDES, National Pollutant Discharge Elimination System; NRI, National Resource Inventory; PAD, Protected Area Database; PRISM, Parameter-elevation Regressions on Independent Slopes Model; RUSLE, Revised Universal Soil Loss Equation; SSURGO, Soil Survey Geographic (database); STATSGO, State Soil Geographic (database); STATSGO2, U.S. General Soil Map (database); TIGER, Topologically Integrated Geographic Encoding and Referencing (system); TRI, Toxic Release Inventory; USEPA, U.S. Environmental Protection Agency; m, meters; km, kilometers; ~, approximately; –, not applicable]

Method used by NAWQA	General category	Characteristic	Description of the national thematic dataset	Data format, and scale or resolution of dataset used by NAWQA	Original data format, scale or resolution, if different	Data references and miscellaneous notes
Simple overlay	Climate	Mean of monthly maximum and minimum number of days of measurable precipitation: 1961–1990	Monthly maximum and minimum number of days of measurable precipitation, derived from 30-years of record, 1961–1990, compiled by the PRISM Climate Group	2-km resolution rasters	4-km resolution rasters	From the PRISM Climate Group (2010), developed using methods by Daly and others (2002). The original 4-km resolution rasters were resampled to 2-km resolution by Climate Source, Inc. (2006; Ryan Hill, Utah State University, personal commun., November 2006).
Simple overlay	Climate	Mean of average day of the year of first freeze: 1961–1990	Average day of the year (1–365) of first freeze, derived from 30-years of record, 1961–1990, compiled by the PRISM Climate Group	2-km resolution raster	4-km resolution raster	From the PRISM Climate Group (2010), developed using methods by Daly and others (2002). The original 4-km resolution raster was resampled to 2-km resolution by Climate Source, Inc. (2006; Ryan Hill, Utah State University, personal commun., November 2006).
Simple overlay	Climate	Mean of average day of the year of last freeze: 1961–1990	Average day of the year (1–365) of last freeze, derived from 30 years of record, 1961–1990, compiled by the PRISM Climate Group	2-km resolution raster	4-km resolution raster	From the PRISM Climate Group (2010), developed using methods by Daly and others (2002). The original 4-km resolution raster was resampled to 2-km resolution by Climate Source, Inc. (2006; Ryan Hill, Utah State University, personal commun., November 2006).
Simple overlay	Climate	Mean of average annual potential evapotranspiration: 1961–1990	Average annual potential evaporation for the period 1961–1990	1-km resolution raster	0.04167 decimal degrees raster	Derived from temperature data developed by the PRISM Climate Group (2010) used with Hamon's (1961) equation for potential evaporation to calculate average annual potential evaporation for 1961–1990. PRISM's original raster was spatially interpolated to a 1-km resolution raster (David M. Wolock, U.S. Geological Survey, written commun., 2001).
Simple overlay	Climate	Mean of average annual potential evapotranspiration: 1971–2000	Average annual potential evaporation for the period 1971–2000	1-km resolution raster	0.04167 decimal degrees raster	Derived from temperature data developed by the PRISM Climate Group (2010) used with Hamon's (1961) equation for potential evaporation to calculate average annual potential evaporation for 1971–2000. PRISM's original raster was spatially interpolated to a 1-km resolution raster (David M. Wolock, U.S. Geological Survey, written commun., 2001).

Appendix A. The national GIS thematic datasets used by the National Water-Quality Assessment (NAWQA) program to characterize sampling sites for streams and groundwater.—Continued

[**Abbreviations**: AVHRR, Advanced Very High Resolution Radiometer; CDL, Cropland Data Layer; CTIC, Conservation Technology Information Center; DEM, Digital Elevation Model; GIS, geographic information system; LULC, Land Use and Land Cover; MSA, Metropolitan Statistical Areas; NADP, National Atmospheric Deposition Program; NAWQA, National Water-Quality Assessment (program); NCDC, National Climatic Data Center; NED, National Elevation Data; NHDPlus, National Hydrography Dataset Plus; NLCD 2001, National Land Cover Database 2001; NLCD 2006, National Land Cover Database 2006; NLCDe 92, Enhanced version of the National Land Cover Data 1992; NLCDeP, Enhanced National Land Cover Data 1992 revised with 1990 and 2000 population data; NPDES, National Pollutant Discharge Elimination System; NRI, National Resource Inventory; PAD, Protected Area Database; PRISM, Parameter-elevation Regressions on Independent Slopes Model; RUSLE, Revised Universal Soil Loss Equation; SSURGO, Soil Survey Geographic (database); STATSGO, State Soil Geographic (database); STATSGO2, U.S. General Soil Map (database); TIGER, Topologically Integrated Geographic Encoding and Referencing (system); TRI, Toxic Release Inventory; USEPA, U.S. Environmental Protection Agency; m, meters; km, kilometers; ~, approximately; –, not applicable]

Method used by NAWQA	General category	Characteristic	Description of the national thematic dataset	Data format, and scale or resolution of dataset used by NAWQA	Original data format, scale or resolution, if different	Data references and miscellaneous notes
Simple overlay	Climate	Mean of average annual precipitation: 1971–2000	Average annual precipitation for the period 1971–2000, compiled by the PRISM Climate Group	1-km resolution raster	30-arc seconds (~800-m) resolution raster	From the PRISM Climate Group (2010), developed using methods by Daly and others (2002). PRISM's original raster was spatially interpolated to a 1-km resolution raster (David M Wolock, U.S. Geological Survey, written commun., 2005).
Simple overlay	Climate	Mean of average annual air temperature: 1971–2000	Average annual air temperature for the period 1971–2000, compiled by the PRISM Climate Group	1-km resolution raster	30-arc seconds (~800-m) resolution raster	From the PRISM Climate Group (2010), developed using methods by Daly and others (2002). PRISM's original raster was spatially interpolated to a 1-km resolution raster (David M Wolock, U.S. Geological Survey, written commun., 2005).
Simple overlay	Climate	Mean of average annual precipitation: 1980–1997	Average annual precipitation for the period 1980–1997, from the Daymet database	1-km resolution raster	Points	From ground-based meterological stations and adjusted elevation in the Daymet database (Thorton and Running, 1999) developed at the University of Montana, by the Numerical Terradynamic Simulation Group (2005).
Simple overlay	Climate	Mean of monthly precipitation: 1980–1997	Average precipitation for each month, for the period 1980–1997, from the Daymet database	1-km resolution rasters	Points	From ground-based meterological stations and adjusted elevation in the Daymet database (Thorton and Running, 1999) developed at the University of Montana, by the Numerical Terradynamic Simulation Group (2005). There is a single raster for each month for the 18-year period from 1980–1997 for a total of 12 rasters.
Simple overlay	Climate	Mean of average annual temperature: 1980–1997	Average annual temperature for the period 1980–1997, from the Daymet database	1-km resolution raster	Points	From ground-based meterological stations and adjusted elevation in the Daymet database (Thorton and Running, 1999) developed at the University of Montana, by the Numerical Terradynamic Simulation Group (2005).
Simple overlay	Climate	Mean of monthly temperature: 1980–1997	Average temperature for each month, for the period 1980–1997, from the Daymet database	1-km resolution rasters	Points	From ground-based meterological stations and adjusted elevation in the Daymet database (Thorton and Running, 1999) developed at the University of Montana, Numerical Terradynamic Simulation Group (2005). There is a single raster for each month for the 18-year period from 1980–1997 for a total of 12 rasters.
Simple overlay	Geology	Percent bedrock and surficial geology	Bedrock and surficial geology	1-km resolution raster	Between 1:10,000,000- and 1:7,500,000-scale polygons	From vector polygons of bedrock and surficial geology (Reed and Bush, 2005) that were rasterized to a 1-km resolution raster.
Simple overlay	Geology	Percent bedrock geology	Bedrock geology	30-m resolution raster	1:2,500,000-scale polygons	From vector polygons of bedrock geology (King and Beikman, 1974a,b), compiled digitally by Schruben and others (1998), and rasterized to a 30-m resolution raster.

Appendix A. The national GIS thematic datasets used by the National Water-Quality Assessment (NAWQA) program to characterize sampling sites for streams and groundwater.—Continued

[**Abbreviations**: AVHRR, Advanced Very High Resolution Radiometer; CDL, Cropland Data Layer; CTIC, Conservation Technology Information Center; DEM, Digital Elevation Model; GIS, geographic information system; LULC, Land Use and Land Cover; MSA, Metropolitan Statistical Areas; NADP, National Atmospheric Deposition Program; NAWQA, National Water-Quality Assessment (program); NCDC, National Climatic Data Center; NED, National Elevation Data; NHDPlus, National Hydrography Dataset Plus; NLCD 2001, National Land Cover Database 2001; NLCD 2006, National Land Cover Database 2006; NLCDe 92, Enhanced version of the National Land Cover Data 1992; NLCDeP, Enhanced National Land Cover Data 1992 revised with 1990 and 2000 population data; NPDES, National Pollutant Discharge Elimination System; NRI, National Resource Inventory; PAD, Protected Area Database; PRISM, Parameter-elevation Regressions on Independent Slopes Model; RUSLE, Revised Universal Soil Loss Equation; SSURGO, Soil Survey Geographic (database); STATSGO, State Soil Geographic (database); STATSGO2, U.S. General Soil Map (database); TIGER, Topologically Integrated Geographic Encoding and Referencing (system); TRI, Toxic Release Inventory; USEPA, U.S. Environmental Protection Agency; m, meters; km, kilometers; ~, approximately; –, not applicable]

Method used by NAWQA	General category	Characteristic	Description of the national thematic dataset	Data format, and scale or resolution of dataset used by NAWQA	Original data format, scale or resolution, if different	Data references and miscellaneous notes
Simple overlay	Geology	Percent surficial geology	Surficial geology	30-m resolution raster	1:7,500,000-scale polygons	From vector polygons of surficial geology (Hunt, 1979) published as part of USGS National Atlas map series, compiled digitally by Clawges and Price (1999a), and converted to a 30-m resolution raster.
Simple overlay	Geology	Percent surficial quaternary sediments	Surficial quaternary sediments[17]	1:1,000,000-scale polygons	–	From Soller and Packard (1998).
Simple overlay	Hydrology	Stream density	Streams	1:100,000-scale lines	–	From NHDPlus hydrography (U.S. Geological Survey and U.S. Environmental Protection Agency, 2010; U.S. Environmental Protection Agency, 2008). Stream density was calculated as the total length of streams in the study area divided by the area of the study area.
Simple overlay	Hydrology	Percent of stream-line characteristics[18]	Streams	1:100,000-scale lines	–	From NHDPlus hydrography (U.S. Geological Survey and U.S. Environmental Protection Agency, 2010; U.S. Environmental Protection Agency, 2008).
Simple overlay	Hydrology	Stream segment[19] characteristic: sinuosity	Streams	1:100,000-scale lines	–	From NHDPlus hydrography (U.S. Geological Survey and U.S. Environmental Protection Agency, 2010; U.S. Environmental Protection Agency, 2008). Segment sinuosity was calculated as the curvilinear length of the segment stream line divided by the straight-line distance between the end points of the line.
Simple overlay	Hydrology	Stream-segment[19] characteristic: gradient	Streams	1:100,000-scale lines	–	From NHDPlus hydrography (U.S. Geological Survey and U.S. Environmental Protection Agency, 2010; U.S. Environmental Protection Agency, 2008), and elevation from the NED (U.S. Geological Survey, 1999b, 2003). Segment gradient was calculated from the elevation difference between the end points of the line divided by the curvilinear length.

Appendix A. The national GIS thematic datasets used by the National Water-Quality Assessment (NAWQA) program to characterize sampling sites for streams and groundwater.—Continued

[**Abbreviations**: AVHRR, Advanced Very High Resolution Radiometer; CDL, Cropland Data Layer; CTIC, Conservation Technology Information Center; DEM, Digital Elevation Model; GIS, geographic information system; LULC, Land Use and Land Cover; MSA, Metropolitan Statistical Areas; NADP, National Atmospheric Deposition Program; NAWQA, National Water-Quality Assessment (program); NCDC, National Climatic Data Center; NED, National Elevation Data; NHDPlus, National Hydrography Dataset Plus; NLCD 2001, National Land Cover Database 2001; NLCD 2006, National Land Cover Database 2006; NLCDe 92, Enhanced version of the National Land Cover Data 1992; NLCDeP, Enhanced National Land Cover Data 1992 revised with 1990 and 2000 population data; NPDES, National Pollutant Discharge Elimination System; NRL, National Resource Inventory; PAD, Protected Area Database; PRISM, Parameter-elevation Regressions on Independent Slopes Model; RUSLE, Revised Universal Soil Loss Equation; SSURGO, Soil Survey Geographic (database); STATSGO, State Soil Geographic (database); STATSGO2, U.S. General Soil Map (database); TIGER, Topologically Integrated Geographic Encoding and Referencing (system); TRI, Toxic Release Inventory; USEPA, U.S. Environmental Protection Agency; m, meters; km, kilometers; ~, approximately; —, not applicable]

Method used by NAWQA	General category	Characteristic	Description of the national thematic dataset	Data format, and scale or resolution of dataset used by NAWQA	Original data format, scale or resolution, if different	Data references and miscellaneous notes
Simple overlay	Hydrology	Stream-segment[19] characteristic: mean distance to roads	Roads	1:100,000-scale lines	—	From the TIGER/Line roads database (U.S. Census Bureau, 2011) obtained from GeoLytics, Inc. (2001b) and the NHDPlus hydrography (U.S. Geological Survey and U.S. Environmental Protection Agency, 2010; U.S. Environmental Protection Agency, 2008). The mean distance to roads was calculated by converting segment stream lines to 30-m point locations, then determining the mean of the distance of each point to the nearest road.
Simple overlay	Hydrology	Stream segment[19] characteristics: percent land cover	Streams and land cover	Streams: 1:100,000-scale lines; Land cover: 30-m resolution raster	—	From the NHDPlus hydrography (U.S. Geological Survey and U.S. Environmental Protection Agency, 2010) and the NLCDe 92[3] (Nakagaki and others 2007). Segment percent land cover was calculated within a 100-m buffer on each side of the segment stream line.
Simple overlay	Hydrology	Mean of 30-year average-annual runoff: 1951–1980	Thirty-year average annual runoff computed from streamflow data collected during 1951–1980	1-km resolution raster	1:7,500,000-scale lines	From Gebert and others (1987).
Simple overlay	Hydrology	Mean of 30-year average annual runoff: 1971–2000	Thirty-year average annual runoff (streamflow per unit area) from 1971 through 2000	1-km resolution raster	Variable-sized basins and hydrologic cataloging units at variable resolutions	Developed by David M. Wolock (U.S. Geological Survey, written commun., 2012) by combining historical streamflow data (U.S. Geological Survey, 2010) collected from 1971–2000 and hydrologic cataloging units in the conterminous U.S. (Steeves and Nebert, 1994) following the approach of Krug and others (1989).
Simple overlay	Hydrology	Mean annual runoff: 1990–2011	Annual time series of runoff (streamflow per unit area) from 1990 through 2011	1-km resolution rasters	Variable-sized basins and hydrologic cataloging units at variable resolutions	Developed by David M. Wolock (U.S. Geological Survey, written commun., 2005) by combining streamflow data (U.S. Geological Survey, 2012) collected from 1990–2011 and hydrologic cataloging units in the conterminous U.S. (Steeves and Nebert, 1994) following the approach of Krug and others (1989). A single raster was generated for each year.
Simple overlay	Hydrology	Mean base-flow index	Base-flow index, expressed as a percentage	1-km resolution raster	—	From Wolock (2003a). Base-flow index is defined as the component of stream flow that can be attributed to ground-water discharge into streams, and is the ratio of base flow to total flow (Wolock, 2003a).

Appendix A. The national GIS thematic datasets used by the National Water-Quality Assessment (NAWQA) program to characterize sampling sites for streams and groundwater.—Continued

[Abbreviations: AVHRR, Advanced Very High Resolution Radiometer; CDL, Cropland Data Layer; CTIC, Conservation Technology Information Center; DEM, Digital Elevation Model; GIS, geographic information system; LULC, Land Use and Land Cover; MSA, Metropolitan Statistical Areas; NADP, National Atmospheric Deposition Program; NAWQA, National Water-Quality Assessment (program); NCDC, National Climatic Data Center; NED, National Elevation Data; NHDPlus, National Hydrography Dataset Plus; NLCD 2001, National Land Cover Database 2001; NLCD 2006, National Land Cover Database 2006; NLCDe 92, Enhanced version of the National Land Cover Data 1992; NLCDeP, Enhanced National Land Cover Data 1992 revised with 1990 and 2000 population data; NPDES, National Pollutant Discharge Elimination System; NRI, National Resource Inventory; PAD, Protected Area Database; PRISM, Parameter-elevation Regressions on Independent Slopes Model; RUSLE, Revised Universal Soil Loss Equation; SSURGO, Soil Survey Geographic (database); STATSGO, State Soil Geographic (database); STATSGO2, U.S. General Soil Map (database); TIGER, Topologically Integrated Geographic Encoding and Referencing (system); TRI, Toxic Release Inventory; USEPA, U.S. Environmental Protection Agency; m, meters; km, kilometers; ~, approximately; –, not applicable]

Method used by NAWQA	General category	Characteristic	Description of the national thematic dataset	Data format, and scale or resolution of dataset used by NAWQA	Original data format, scale or resolution, if different	Data references and miscellaneous notes
Simple overlay	Hydrology	Mean of average percentage of Dunne overland flow	Average percentage of saturation overland flow in total streamflow	5-km resolution raster	–	From Wolock (2003b). The average percentage of saturation overland flow in total streamflow is simulated in a watershed model, "TOPMODEL" (Beven and Kirkby, 1979) as precipitation that falls on saturated land-surface areas and enters the stream channel (Wolock, 2003b).
Simple overlay	Hydrology	Mean ground-water recharge	Index of mean annual natural groundwater recharge from precipitation	1-km resolution raster	–	From Wolock (2003c).
Simple overlay	Hydrology	Mean of average percentage of Horton overland flow	Average percentage of infiltration-excess overland flow in total streamflow	5-km resolution raster	–	The average percentage of infiltration-excess overland flow in total streamflow is simulated in a watershed model "TOPMODEL" (Beven and Kirkby, 1979) as precipitation that exceeds the infiltration capacity of the soil and enters the stream channel (Wolock, 2003d).
Simple overlay	Hydrology	Mean subsurface flow contact time	Index of subsurface flow contact time, expressed in days (residence time)	5-km resolution raster	–	From Wolock (2003d) Derived from watershed model equations that used mapped STATSGO soil characteristics (U.S Department of Agriculture, 1994) from Wolock (1997) and DEM data (U.S. Geological Survey, 1993, 2001) at the 1:250,000-scale (David M. Wolock, U.S. Geological Survey, written commun., 2001).
Simple overlay	Hydrology	Percent principal aquifer	Productive aquifers	1:7,500,000-scale polygons	–	From a national map of productive aquifers (U.S. Geological Survey, 1970), compiled digitally by Clawges and Price (1999b).
Simple overlay	Hydrology	Mean topographic wetness index	Topographic wetness index	1-km resolution raster	90-m resolution raster	Developed by David M. Wolock (U.S. Geological Survey, written commun., 2001) Topographic index values were derived from DEM data (U.S. Geological Survey, 1993, 2001; Verdin and Greenlee, 1996) at the 1:250,000-scale. The topographic wetness index is an indicator for potential soil moisture based on slope of the landscape and subsurface flow. The single flow direction algorithm was used to compute this version of the topographic index, which is calculated from ln (a/tan B) where "a" is the upslope area per unit contour length and tan B is the slope gradient (Wolock and McCabe, 1995).

Appendix A.   The national GIS thematic datasets used by the National Water-Quality Assessment (NAWQA) program to characterize sampling sites for streams and groundwater.—Continued

[Abbreviations: AVHRR, Advanced Very High Resolution Radiometer; CDL, Cropland Data Layer; CTIC, Conservation Technology Information Center; DEM, Digital Elevation Model; GIS, geographic information system; LULC, Land Use and Land Cover; MSA, Metropolitan Statistical Areas; NADP, National Atmospheric Deposition Program; NAWQA, National Water-Quality Assessment (program); NCDC, National Climatic Data Center; NED, National Elevation Data; NHDPlus, National Hydrography Dataset Plus; NLCD 2001, National Land Cover Database 2001; NLCD 2006, National Land Cover Database 2006; NLCDe 92, Enhanced version of the National Land Cover Data 1992; NLCDeP, Enhanced National Land Cover Data 1992 revised with 1990 and 2000 population data; NPDES, National Pollutant Discharge Elimination System; NRI, National Resource Inventory; PAD, Protected Area Database; PRISM, Parameter-elevation Regressions on Independent Slopes Model; RUSLE, Revised Universal Soil Loss Equation; SSURGO, Soil Survey Geographic (database); STATSGO, State Soil Geographic (database); STATSGO2, U.S. General Soil Map (database); TIGER, Topologically Integrated Geographic Encoding and Referencing (system); TRI, Toxic Release Inventory; USEPA, U.S. Environmental Protection Agency; m, meters; km, kilometers; ~, approximately; --, not applicable]

Method used by NAWQA	General category	Characteristic	Description of the national thematic dataset	Data format, and scale or resolution of dataset used by NAWQA	Original data format, scale or resolution, if different	Data references and miscellaneous notes
Land-cover weighted areal inter-polation	Land and water management	Area of irrigated land and conservation management practices[20]: 1992	Areas of irrigated land and conservation management practices on agricultural land	Irrigation and conservation management practices data: tabular by county; county boundaries: 30-m resolution raster; agricultural land: 30-m resolution raster	County boundaries: 1:100,000-scale polygons and 1:70,000-scale lines	From the 1992 NRI (U.S. Department of Agriculture, 1995) compiled at the county level (Michael E. Wieczorek, U.S. Geological Survey written commun., 2004), then linked to mapped 1990 county boundaries[5] and mapped agricultural land from the NLCDe 92[3,6] (Nakagaki and others, 2007).
Land-cover weighted areal inter-polation	Land and water management	Area of irrigated land and conservation management practices[21]: 1997	Areas of irrigated land and conservation management practices on agricultural land	Irrigation and conservation management practices data: tabular by county; county boundaries: 30-m resolution raster; agricultural land: 30-m resolution raster	County boundaries: 1:100,000-scale polygons and 1:70,000-scale lines	From the 1997 NRI (U.S. Department of Agriculture, 2000a,b) compiled at the county level (Michael E. Wieczorek, U.S. Geological Survey written commun., 2004) then linked to mapped 1990 county boundaries[5] and mapped agricultural land from the NLCDe 92[3,6] (Nakagaki and others, 2007).
Land-cover weighted areal inter-polation	Land and water management	Area of tilled land: annually from 1990 through 2001	Areas of conservation tillage practices (no till, ridge till, mulch till, reduced till, and conventional or intense till)	Tillage data: tabular by county; county boundaries: 30-m resolution rasters; agricultural land: 30-m resolution rasters	County boundaries: 1:100,000-scale polygons and 1:70,000-scale lines	From CTIC tillage data assembled at the county level (Nancy T. Baker, U.S. Geological Survey, written commun., 2010). Tillage data from 1990 through 1997 were linked to mapped 1990 county boundaries[5] and mapped agricultural land from the NLCDe 92[3,6] (Nakagaki and others, 2007); and tillage data from 1998–2001 were linked to mapped 2001 county boundaries[9] and mapped agricultural land[8] from the NLCD 2001[7] (U.S. Geological Survey, 2007a; Homer and others, 2007; LaMotte 2008,a,b,c,d). CTIC's tillage data have been aggregated by 8-digit hydrologic units (Baker, 2011).

Appendix A. The national GIS thematic datasets used by the National Water-Quality Assessment (NAWQA) program to characterize sampling sites for streams and groundwater.—Continued

[Abbreviations: AVHRR, Advanced Very High Resolution Radiometer; CDL, Cropland Data Layer; CTIC, Conservation Technology Information Center; DEM, Digital Elevation Model; GIS, geographic information system; LULC, Land Use and Land Cover; MSA, Metropolitan Statistical Areas; NADP, National Atmospheric Deposition Program; NAWQA, National Water-Quality Assessment (program); NCDC, National Climatic Data Center; NED, National Elevation Data; NHDPlus, National Hydrography Dataset Plus; NLCD 2001, National Land Cover Database 2001; NLCD 2006, National Land Cover Database 2006; NLCDe 92, Enhanced version of the National Land Cover Data 1992; NLCDeP, Enhanced National Land Cover Data 1992 revised with 1990 and 2000 population data; NPDES, National Pollutant Discharge Elimination System; NRI, National Resource Inventory; PAD, Protected Area Database; PRISM, Parameter-elevation Regressions on Independent Slopes Model; RUSLE, Revised Universal Soil Loss Equation; SSURGO, Soil Survey Geographic (database); STATSGO, State Soil Geographic (database); STATSGO2, U.S. General Soil Map (database); TIGER, Topologically Integrated Geographic Encoding and Referencing (system); TRI, Toxic Release Inventory; USEPA, U.S. Environmental Protection Agency; m, meters; km, kilometers; ~, approximately; –, not applicable]

Method used by NAWQA	General category	Characteristic	Description of the national thematic dataset	Data format, and scale or resolution of dataset used by NAWQA	Original data format, scale or resolution, if different	Data references and miscellaneous notes
Land-cover weighted areal interpolation	Land and water management	Volume of water withdrawn for irrigation and area of irrigated land: 1995	Volume of groundwater and surface-water withdrawals for irrigation, total irrigation consumptive use, total irrigated acres, and reclaimed wastewater - all for 1995	Water withdrawn for irrigation and area of irrigated land: tabular by county; agricultural land: 30-m resolution raster; county boundaries: 30-m resolution raster	County boundaries: 1:100,000-scale polygons and 1:70,000-scale lines	From Solley and others (1998), combined with mapped 1990 county boundaries[5] and mapped agricultural land from the NLCDe 92[3,6] (Nakagaki and others, 2007).
Land cover-weighted areal interpolation	Land and water management	Area of crops: 1992, 1997, 2002, 2007	Harvested acreages of row crops, orchard crops, and pasture/hay crops	Harvested crop areas: tabular by county; agricultural land: 30-m resolution raster; county boundaries: 30-m resolution raster	County boundaries: 1:100,000-scale polygons and 1:70,000-scale lines	Derived from county-level harvested crop acreages from the 2007 Census of Agriculture (U.S. Department of Agriculture, 2009), 2002 Census of Agriculture (U.S. Department of Agriculture, 2004), 1997 Census of Agriculture (U.S. Department of Agriculture, 1999), and the 1992 Census of Agriculture (U.S. Department of Commerce, 1995). For crop data years 1992 and 1997, the NLCDe 92[3] (Nakagaki and others, 2007) was used for the source for mapped agricultural land[6] combined with mapped 1990 county boundaries[5]; for crop data years 2002 and 2007, the NLCD 2001[7] (U.S. Geological Survey, 2007a; Homer and others, 2007; LaMotte, 2008a,b,c,d) was used for the source for mapped agricultural land[8] combined with mapped 2001 county boundaries[9]. For some applications, the crop information by study area is further categorized into major crop groups or orchard-and-vineyard crop groups, using methods developed by Gilliom and Thelin (1997).
Simple overlay	Land cover	Percent land by crop	Crop-specific land cover, annually from 2008 through 2011	30-m resolution rasters	–	From the CDL, developed by the National Agricultural Statistics Service (U.S. Department of Agriculture, 2010a; Johnson and Mueller, 2010).
Simple overlay	Land cover	Percent land cover: ~2006	Land cover for the mid 2000s	30-m resolution raster	–	From the NLCD 2006 (Fry and others, 2011; U.S. Geological Survey, 2011).
Simple overlay	Land cover	Percent land cover: ~2001	Land cover for the late 1990s to early 2000s	30-m resolution raster	–	From the NLCD 2001[7] (U.S. Geological Survey, 2007a; Homer and others, 2007; LaMotte 2008a,b,c,d).

Appendix A. The national GIS thematic datasets used by the National Water-Quality Assessment (NAWQA) program to characterize sampling sites for streams and groundwater.—Continued

[**Abbreviations**: AVHRR, Advanced Very High Resolution Radiometer; CDL, Cropland Data Layer; CTIC, Conservation Technology Information Center; DEM, Digital Elevation Model; GIS, geographic information system; LULC, Land Use and Land Cover; MSA, Metropolitan Statistical Areas; NADP, National Atmospheric Deposition Program; NAWQA, National Water-Quality Assessment (program); NCDC, National Climatic Data Center; NED, National Elevation Data; NHDPlus, National Hydrography Dataset Plus; NLCD 2001, National Land Cover Database 2001; NLCD 2006, National Land Cover Database 2006; NLCDe 92, Enhanced version of the National Land Cover Data 1992; NLCDeP, Enhanced version of the National Land Cover Data 1992 revised with 1990 and 2000 population data; NPDES, National Pollutant Discharge Elimination System; NRI, National Resource Inventory; PAD, Protected Area Database; PRISM, Parameter-elevation Regressions on Independent Slopes Model; RUSLE, Revised Universal Soil Loss Equation; SSURGO, Soil Survey Geographic (database); STATSGO, State Soil Geographic (database); STATSGO2, U.S. General Soil Map (database); TIGER, Topologically Integrated Geographic Encoding and Referencing (system); TRI, Toxic Release Inventory; USEPA, U.S. Environmental Protection Agency; m, meters; km, kilometers; ~, approximately; –, not applicable]

Method used by NAWQA	General category	Characteristic	Description of the national thematic dataset	Data format, and scale or resolution of dataset used by NAWQA	Original data format, scale or resolution, if different	Data references and miscellaneous notes
Simple overlay	Land cover	Percent land cover change: 1992 vs. 2001	Land cover change comparison from 1992 and 2001, at the Anderson Level I classification scale (Anderson and others, 1976)	30-m resolution raster	–	From the National Land Cover Database 1992–2001 Retrofit Land Cover Change Product (Fry and others, 2009, U.S. Geological Survey, 2007b).
Simple overlay	Land cover	Percent land cover: ~1992	Land cover for the early 1990s	30-m resolution raster	–	From the NLCDe 92[3] (Nakagaki and others, 2007).
Simple overlay	Land cover	Percent land cover: 1990s revised for 2000 residential urban	Land cover for the 1990s revised for 2000 residential urban	30-m resolution raster	–	From the NLCDeP[12] (Hitt, 2008).
Simple overlay	Land cover	Percent land use and land cover: 1970s	Land use and land cover for the 1970s	100-m resolution raster	1:250,000-scale polygons	From the LULC[1] dataset (U.S. Geological Survey, 1990, 1998).
Simple overlay	Land cover	Percent land use and land cover: 1970s, revised for new residential development using 1990 population density	Land use and land cover for the 1970s, revised for new (1990s) residential development	100-m resolution raster	1:250,000-scale polygons	Derived by combining the 1970s LULC dataset[1] (U.S. Geological Survey, 1990, 1998) and 1990 population density (U.S. Census Bureau, 1992) using methods developed by Hitt (1994).
Simple overlay	Other physical parameters	Mean aspect, northness and eastness	Elevation	100-m resolution raster	30-m resolution raster	From the 30-m resolution NED (U.S. Geological Survey, 1999b, 2003), which was resampled to 100-m resolution. The 100-m resolution NED raster was used to generate a raster representing the aspect of the study area from which the mean aspect was calculated. The mean aspect was computed by taking the tangent of the sum of the sine of the angle (in radians) divided by the sum of the cosine of the angle (in radians).

Appendix A. The national GIS thematic datasets used by the National Water-Quality Assessment (NAWQA) program to characterize sampling sites for streams and groundwater.—Continued

[Abbreviations: AVHRR, Advanced Very High Resolution Radiometer; CDL, Cropland Data Layer; CTIC, Conservation Technology Information Center; DEM, Digital Elevation Model; GIS, geographic information system; LULC, Land Use and Land Cover; MSA, Metropolitan Statistical Areas; NADP, National Atmospheric Deposition Program; NAWQA, National Water-Quality Assessment (program); NCDC, National Climatic Data Center; NED, National Elevation Data; NHDPlus, National Hydrography Dataset Plus; NLCD 2001, National Land Cover Database 2001; NLCD 2006, National Land Cover Database 2006; NLCDe 92, Enhanced version of the National Land Cover Data 1992; NLCDeP, Enhanced National Land Cover Data 1992 revised with 1990 and 2000 population data; NPDES, National Pollutant Discharge Elimination System; NRI, National Resource Inventory; PAD, Protected Area Database; PRISM, Parameter-elevation Regressions on Independent Slopes Model; RUSLE, Revised Universal Soil Loss Equation; SSURGO, Soil Survey Geographic (database); STATSGO, State Soil Geographic (database); STATSGO2, U.S. General Soil Map (database); TIGER, Topologically Integrated Geographic Encoding and Referencing (system); TRI, Toxic Release Inventory; USEPA, U.S. Environmental Protection Agency; m, meters; km, kilometers; ~, approximately; –, not applicable]

Method used by NAWQA	General category	Characteristic	Description of the national thematic dataset	Data format, and scale or resolution of dataset used by NAWQA	Original data format, scale or resolution, if different	Data references and miscellaneous notes
Simple overlay	Other physical parameters	Mean percent slope	Elevation	100-m resolution raster	30-m resolution raster	From the 30-m resolution NED (U.S. Geological Survey, 1999b, 2003), which was resampled to 100-m resolution. The 100-m resolution NED raster was used to generate a raster of the slope of the study area from which the mean values of the grid cells within the study area was determined.
Simple overlay	Other physical parameters	Mean elevation of study area and elevation at site	Elevation	100-m resolution raster	30-m resolution raster	From the 30-m resolution NED (U.S. Geological Survey, 1999b, 2003), resampled to 100-m resolution.
Simple overlay	Other physical parameters	Land cover fragmentation	Land cover	30-m resolution raster	–	Derived from the method developed by Ritters and others (2000). The undeveloped land is based on selected land classifications from the NLCDe 92[22] (Nakagaki and others, 2007). Higher numbers equal more fragmentation by developed land cover. Using a 3x3 processing window, the calculation was 100 minus percent watershed in "Ritters class 1".
Simple overlay	Regions	Percent in protected areas	Protected areas in national parks, wilderness areas, and wildlife refuges	1:100,000-scale polygons	–	From the Conservation Biology Institute's (2006) PAD.
Simple overlay	Regions	Percent level III ecoregions	Level III ecoregions	30-m resolution raster	1:7,500,000-scale polygons	From vector polygons of level III ecoregions (Omernik, 1987) and rasterized to 30-m resolution; GIS files of ecoregions are available from the U.S. Environmental Protection Agency (2010b).
Simple overlay	Regions	Percent level IV ecoregions	Level IV ecoregions	30-m resolution raster	1:7,500,000-scale polygons	From vector polygons of level IV ecoregions (Omernik, 1987) and rasterized to 30-m resolution; GIS files of ecoregions are available from the U.S. Environmental Protection Agency (2010b).
Simple overlay	Regions	Percent farm resource regions	Farm resource regions	100-m resolution raster	1:2,000,000-scale polygons	From vector polygons of clusters of counties that make up the Economic Research Service farm resource regions based on climate, topography, soil types, and dominant agricultural activities (U.S. Department of Agriculture, 2000c). Rasterized to 100-m resolution.
Simple overlay	Regions	Percent hydrologic landscape regions	Hydrologic landscape regions	100-m resolution raster	–	From hydrologic landscape regions, which consist of aggregates of watersheds on basis of similarities in land-surface form, geologic texture, and climate characteristics with boundaries (Wolock, 2003e). Based on the 100-m resolution version (David M. Wolock, U.S. Geological Survey, written commun., 2004).

Appendix A. The national GIS thematic datasets used by the National Water-Quality Assessment (NAWQA) program to characterize sampling sites for streams and groundwater.—Continued

[**Abbreviations**: AVHRR, Advanced Very High Resolution Radiometer; CDL, Cropland Data Layer; CTIC, Conservation Technology Information Center; DEM, Digital Elevation Model; GIS, geographic information system; LULC, Land Use and Land Cover; MSA, Metropolitan Statistical Areas; NADP, National Atmospheric Deposition Program; NAWQA, National Water-Quality Assessment (program); NCDC, National Climatic Data Center; NED, National Elevation Data; NHDPlus, National Hydrography Dataset Plus; NLCD 2001, National Land Cover Database 2001; NLCD 2006, National Land Cover Database 2006; NLCDe 92, Enhanced version of the National Land Cover Data 1992; NLCDeP, Enhanced National Land Cover Data 1992 revised with 1990 and 2000 population data; NPDES, National Pollutant Discharge Elimination System; NRI, National Resource Inventory; PAD, Protected Area Database; PRISM, Parameter-elevation Regressions on Independent Slopes Model; RUSLE, Revised Universal Soil Loss Equation; SSURGO, Soil Survey Geographic (database); STATSGO, State Soil Geographic (database); STATSGO2, U.S. General Soil Map (database); TIGER, Topologically Integrated Geographic Encoding and Referencing (system); TRI, Toxic Release Inventory; USEPA, U.S. Environmental Protection Agency; m, meters; km, kilometers; ~, approximately; –, not applicable]

Method used by NAWQA	General category	Characteristic	Description of the national thematic dataset	Data format, and scale or resolution of dataset used by NAWQA	Original data format, scale or resolution of dataset, if different	Data references and miscellaneous notes
Simple overlay	Regions	Percent nutrient ecoregions	Aggregations of level III ecoregions for national nutrient assessment and management strategy	30-m resolution raster	1:7,500,000-scale polygons	From vector polygons of nutrient ecoregions (U.S. Environmental Protection Agency, 1998; Rohm and others, 2002), acquired from Jeff Comstock (U.S. Environmental Protection Agency, written commun., 2002).
Simple overlay	Regions	Percent physiographic regions	Physiographic regions	30-m resolution raster	1:7,000,000-scale polygons	From vector polygons of physical divisions of the U.S. (Fenneman and Johnson, 1946), rasterized to 30-m resolution.
Simple overlay	Regions	Percent by termite zone in 1970s urban areas	Termite density zones in 1970s urban areas	100-m resolution raster	~1:34,000,000-scale polygons	From vector polygons of mapped termite zones (Beal and others, 1994) combined with 1970s urban areas[23] from the LULC dataset[1] (U.S. Geological Survey, 1990, 1998), and rasterized to 100-m resolution. Weighted termite-urban scores were developed from the relative hazard of subterranean termite infestations and urban areas from the 1970s.
Area-weighted areal interpolation	Soils	Means of various soil characteristics[24], part 1	Weighted averages of selected soil characteristics from the STATSGO database[25], by map unit.	STATSGO map units: 100-m resolution rasters; soil characteristics: tabular by map unit	STATSGO map units: 1:250,000-scale polygons	Derived from STATSGO map-unit vector polygons (U.S. Department of Agriculture, 1994), which were rasterized at the 100-m resolution (David M. Wolock, U.S. Geological Survey, written commun., 2004) and linked to weighted averages of selected STATSGO soil characteristics (Wolock, 1997).
Area-weighted areal interpolation	Soils	Means of various soil characteristics[26], part 2	Weighted averages of selected soil characteristics from the STATSGO database[25], by map unit.	STATSGO map units: 100-m resolution raster; soil characteristics: tabular by map unit	STATSGO map units: 1:250,000-scale polygons	Derived from STATSGO map-unit vector polygons (U.S. Department of Agriculture, 1994), which were rasterized at the 100-m resolution and linked to weighted averages of selected STATSGO soil characteristics (David M. Wolock, U.S. Geological Survey, written commun., 2004).
Area-weighted areal interpolation	Soils	Mean of average soil drainage class (numeric system)	Weighted average of soil drainage class by map unit, from the STATSGO database[24]	STATSGO map units: 100-m resolution raster; soil drainage class; tabular by map unit	STATSGO map units: 1:250,000-scale polygons	Derived from STATSGO map-unit vector polygons (U.S. Department of Agriculture, 1994), which were rasterized at the 100-m resolution (David M. Wolock, U.S. Geological Survey, written commun., 2004) and linked to the weighted averages of STATSGO soil drainage class compiled by Schwarz and Alexander (1995).

Appendix A. The national GIS thematic datasets used by the National Water-Quality Assessment (NAWQA) program to characterize sampling sites for streams and groundwater.—Continued

[Abbreviations: AVHRR, Advanced Very High Resolution Radiometer; CDL, Cropland Data Layer; CTIC, Conservation Technology Information Center; DEM, Digital Elevation Model; GIS, geographic information system; LULC, Land Use and Land Cover; MSA, Metropolitan Statistical Areas; NADP, National Atmospheric Deposition Program; NAWQA, National Water-Quality Assessment (program); NCDC, National Climatic Data Center; NED, National Elevation Data; NHDPlus, National Hydrography Dataset Plus; NLCD 2001, National Land Cover Database 2001; NLCD 2006, National Land Cover Database 2006; NLCDe 92, Enhanced version of the National Land Cover Data 1992; NLCDeP, Enhanced National Land Cover Data 1992 revised with 1990 and 2000 population data; NPDES, National Pollutant Discharge Elimination System; NRI, National Resource Inventory; PAD, Protected Area Database; PRISM, Parameter-elevation Regressions on Independent Slopes Model; RUSLE, Revised Universal Soil Loss Equation; SSURGO, Soil Survey Geographic (database); STATSGO, State Soil Geographic (database); STATSGO2, U.S. General Soil Map (database); TIGER, Topologically Integrated Geographic Encoding and Referencing (system); TRI, Toxic Release Inventory; USEPA, U.S. Environmental Protection Agency; m, meters; km, kilometers; ~, approximately; –, not applicable]

Method used by NAWQA	General category	Characteristic	Description of the national thematic dataset	Data format, and scale or resolution of dataset used by NAWQA	Original data format, scale or resolution, if different	Data references and miscellaneous notes
Simple overlay	Soils	Mean soil erosion	Soil erosion or soil loss per unit area, based on the RUSLE	1-km resolution raster	Multiple sources at various scales	Developed by David M. Wolock (U.S. Geological Survey, written commun., 2005), by applying the RUSLE using various sources including the annual R-factor raster (Daly and others, 2002a,b), the K-factor for the top soil horizon from the STATSGO database (U.S. Department of Agriculture, 1994), the DEM data database (U.S. Geological Survey, 1993), and the NLCDe 92[3] (Nakagaki and others, 2007).
Area-weighted areal interpolation	Soils	Mean percentage soil hydrologic groups[27]	Percentage of soil hydrologic groups by map unit, from enhanced version of the STATSGO database[25]	STATSGO map units: 100-m resolution raster; soil hydrologic group: tabular by map unit	STATSGO map units: 1:250,000-scale polygons	Derived from STATSGO map-unit vector polygons (U.S. Department of Agriculture, 1994) that were rasterized at the 100-m resolution (David M. Wolock, U.S. Geological Survey, written commun., 2004) and linked to the enhanced version of STATSGO soil hydrologic groups in which missing soil hydrologic groups values were populated based on soil characteristics described by Foth and Schafer (1980, Barbara C. Ruddy and William A. Battaglin, U.S. Geological Survey, written commun., 1998).
Area-weighted areal interpolation	Soils	Mean K factor	K factor or "soil erodibility" of the uppermost soil horizon by map unit, from the STATSGO database[25]	STATSGO map units: 100-m resolution raster; K factor: tabular by map unit	STATSGO map units: 1:250,000-scale polygons	Derived from STATSGO map-unit vector polygons (U.S. Department of Agriculture, 1994) that were rasterized at the 100-m resolution and linked to weighted averages of STATSGO K factor (David M. Wolock, U.S. Geological Survey, written commun., 2004).
Area-weighted areal interpolation	Soils	Means of various soil characteristics[27]	Weighted averages of selected soil characteristics from the STATSGO2 database (version 2 of STATSGO, which superseded STATSGO in 2006), by map unit	STATSGO map units: 100-m resolution raster; soil characteristics: tabular by map unit	STATSGO2 map units: 1:250,000-scale polygons	Derived from STATSGO2 map-unit vector polygons (U.S. Department of Agriculture, 2011) that were rasterized at the 100-m resolution and linked to selected STATSGO2 soil characteristics (Michael E. Wieczorek, U.S. Geological Survey, written commun., 2011).

Appendix A.   The national GIS thematic datasets used by the National Water-Quality Assessment (NAWQA) program to characterize sampling sites for streams and groundwater.—Continued

[Abbreviations: AVHRR, Advanced Very High Resolution Radiometer; CDL, Cropland Data Layer; CTIC, Conservation Technology Information Center; DEM, Digital Elevation Model; GIS, geographic information system; LULC, Land Use and Land Cover; MSA, Metropolitan Statistical Areas; NADP, National Atmospheric Deposition Program; NAWQA, National Water-Quality Assessment (program); NCDC, National Climatic Data Center; NED, National Elevation Data; NHDPlus, National Hydrography Dataset Plus; NLCD 2001, National Land Cover Database 2001; NLCD 2006, National Land Cover Database 2006, NLCDe 92, Enhanced version of the National Land Cover Data 1992; NLCDeP, Enhanced National Land Cover Data 1992 revised with 1990 and 2000 population data; NPDES, National Pollutant Discharge Elimination System; NRI, National Resource Inventory; PAD, Protected Area Database; PRISM, Parameter-elevation Regressions on Independent Slopes Model; RUSLE, Revised Universal Soil Loss Equation; SSURGO, Soil Survey Geographic (database); STATSGO, State Soil Geographic (database); STATSGO2, U.S. General Soil Map (database); TIGER, Topologically Integrated Geographic Encoding and Referencing (system); TRI, Toxic Release Inventory; USEPA, U.S. Environmental Protection Agency; m, meters; km, kilometers; ~, approximately; –, not applicable]

Method used by NAWQA	General category	Characteristic	Description of the national thematic dataset	Data format, and scale or resolution of dataset used by NAWQA	Original data format, scale or resolution, if different	Data references and miscellaneous notes
Simple Overlay	Soils	Means of various soil characteristics[28]	Weighted averages of selected soil characteristics from the SSURGO database, by map unit	30-m resolution rasters	SSURGO map units: range from 1:12,000- to 1:63,360-scale polygons; soil characteristics: tabular by map units	Derived from SSURGO map-unit vector polygons (U.S. Department of Agriculture, 2010b) and weighted averages of selected soil characteristics rasterized at the 30-m resolution (Michael E. Wieczorek, U.S. Geological Survey, written commun., 2011). A national raster was developed for each soil characteristic.
Simple overlay	Soils	Mean of average annual R factor	Average annual R-factor or "rainfall erosivity" for the period 1971–2000	1-km resolution raster	0.04167 decimal degrees raster	From Daly and Taylor (2002a,b).

[1] The LULC dataset is based on aerial photographs taken in the 1970s to mid-1980s. The majority of the digital LULC dataset is at the 1:250,000-scale; some areas are at the 1:100,000-scale. It is based on the classification scheme developed by Anderson and others (1976). The LULC dataset used by NAWQA was compiled nationally and enhanced by Price and others (2007).

[2] The LULC classifications that were used to define 1970s urban areas were "residential," "commercial and services," "industrial," "transportation, communications, and utilities," "industrial and commercial complexes," "mixed urban or built-up land," and "other urban or built-up land."

[3] The NLCDe 92 is a combined product of the National Land Cover Dataset 1992 (U.S. Geological Survey, 1999a; Vogelmann and others, 2001), and the LULC dataset (U.S. Geological Survey, 1990, 1998). The National Land Cover Dataset 1992 is based on interpretation of Landsat Thematic Mapper imagery captured from the late 1980s through early 1990s (Vogelmann and others, 2001). The digital data are represented at the 30-m resolution.

[4] The NLCDe 92 classifications that were used to define urban land were "low intensity residential," "high intensity residential," "commercial/industrial/transportation," "LULC residential," "NLCD/LULC forested residential," and "urban/recreational grasses."

[5] The 1990 county boundaries were compiled from 1:100,000-scale county boundaries (U.S. Department of Commerce, 1993) combined with 1:70,000-scale shoreline boundaries (National Oceanic and Atmospheric Administration, 1994; Jo Ann M. Gronberg, U.S. Geological Survey, written commun., 2005).

[6] The NLCDe 92 classifications that were used to define agricultural land were "row crops," "small grains," "fallow," "pasture/hay," "orchards/vineyards/other," and "LULC orchards/vineyards/other."

[7] Version 1 of the NLCD 2001 was released in 2003 and version 2 was released in 2011. Version 2 is directly comparable with the NLCD 2006 and thus is recommended over version 1.

[8] The NLCD 2001 classifications that were used to define agricultural land were "cultivated crops" and "pasture/hay."

[9] The 2001 county boundaries were developed by modifying the existing 1990 county boundaries for the selected county boundaries that had changed in the 1990s. The new boundaries were extracted from the 2000 and 2004 TIGER/Line files (U.S. Census Bureau, 2011; Jo Ann M. Gronberg, U.S. Geological Survey, written commun., 2005).

[10] The LULC classifications that were used to define agricultural land were "cropland and pasture," and "orchards, groves, vineyards, nurseries, and ornamental horticultural areas." The vector polygons of the LULC data (U.S. Geological Survey, 1990, 1998), as enhanced by Price and others (2007), were gridded at the 100-m resolution.

[11] The NLCDe 92 classifications that were used to define urban land were "low intensity residential," "LULC residential," "NLCD/LULC forested residential," and "urban/recreational grasses."

[12] The NLCDeP is a combined product of the NLCDe 92, and 1990 and 2000 population data by census block group geography (Hitt, 2008).

Appendix A. The national GIS thematic datasets used by the National Water-Quality Assessment (NAWQA) program to characterize sampling sites for streams and groundwater.—Continued

[**Abbreviations:** AVHRR, Advanced Very High Resolution Radiometer; CDL, Cropland Data Layer; CTIC, Conservation Technology Information Center; DEM, Digital Elevation Model; GIS, geographic information system; LULC, Land Use and Land Cover; MSA, Metropolitan Statistical Areas; NADP, National Atmospheric Deposition Program; NAWQA, National Water-Quality Assessment (program); NCDC, National Climatic Data Center; NED, National Elevation Data; NHDPlus, National Hydrography Dataset Plus; NLCD 2001, National Land Cover Database 2001; NLCD 2006, National Land Cover Database 2006; NLCDe 92, Enhanced version of the National Land Cover Data 1992; NLCDeP, Enhanced National Land Cover Data 1992 revised with 1990 and 2000 population data; NPDES, National Pollutant Discharge Elimination System; NRI, National Resource Inventory; PAD, Protected Area Database; PRISM, Parameter-elevation Regressions on Independent Slopes Model; RUSLE, Revised Universal Soil Loss Equation; SSURGO, Soil Survey Geographic (database); STATSGO, State Soil Geographic (database); STATSGO2, U.S. General Soil Map (database); TIGER, Topologically Integrated Geographic Encoding and Referencing (system); TRI, Toxic Release Inventory; USEPA, U.S. Environmental Protection Agency: m, meters; km, kilometers; ~, approximately; –, not applicable]

[13] The NLCDeP classifications that were used to define urban land were "low intensity residential," "LULC residential," "NLCD/LULC forested residential," "newly urbanized without trees," "newly urbanized with trees," and "urban/recreational grasses."

[14] The NLCDeP classifications that were used to define agricultural land were "row crops," "small grains," "fallow," "pasture/hay," "orchards/vineyards/other," and "LULC orchards/vineyards/other."

[15] The NLCDe 92 classifications that were used to define agricultural land that represented the distribution of confined animals were "row crops," "small grains," "fallow," and "pasture/hay." The NLCDe 92 classifications that were used to define agricultural land that represented the distribution of unconfined animals were "grasslands/herbaceous," "row crops," "small grains," "fallow," and "pasture/hay."

[16] The NLCDeP classifications that were used to define agricultural land that represented the distribution of confined animals were "row crops," "small grains," "fallow," and "pasture/hay." The NLCDeP classifications that were used to define agricultural land that represented the distribution of unconfined animals were "grasslands/herbaceous," "row crops," "small grains," "fallow," and "pasture/hay."

[17] The map of surficial quaternary sediments (Soller and Packard, 1998) covers the glaciated U.S. east of the Rocky Mountains.

[18] Stream lines are the NHDFlowline features from NHDPlus, which are stream segments that include canals, ditches, artificial paths, pipelines, and coastlines. Stream-line characteristics include maximum Strahler stream order, sinuosity of mainstem stream line, percent of stream length coded as "artificial path," straightline distance of sampling site to nearest canal, and water region.

[19] Stream segment is the length of stream upstream from sampling site as function of drainage area.

[20] Variables include waste management system; irrigation, canal or lateral; conservation tillage system; contour farming; irrigation water conveyance, ditch and canal; irrigation water conveyance, pipeline; irrigation system, tailwater recovery; irrigation water management; irrigation land leveling; terrace farming; subsurface drain; surface drainage, field ditch; surface drainage, main or lateral; no irrigation source; wells as an irrigation source; pond, lake, or reservoir as an irrigation source; stream, ditch or canal as an irrigation source; lagoon, or other waste water (not including tailwater recovery) as an irrigation source; combination of irrigation sources; not irrigated; gravity irrigated; pressure irrigated; and gravity and pressure irrigated.

[21] Variables include contour farming; irrigation system, tailwater recovery; terrace farming; surface drainage, main or lateral; no irrigation source; wells as an irrigation source; pond, lake, or reservoir as an irrigation source; stream, ditch or canal as an irrigation source; lagoon, or other waste water (not including tailwater recovery) as an irrigation source; combination of irrigation sources; not irrigated; gravity irrigated; pressure irrigated; and gravity and pressure irrigated.

[22] The NLCDe 92 classifications that were defined as undeveloped were "open water," "perennial ice/snow," "bare rock/sand/clay," "deciduous forest," "evergreen forest," "mixed forest," "shrubland," "grassland/herbaceous," "woody wetlands," and "emergent herbaceous wetlands."

[23] Mapped zones of relative hazard of subterranean termite infestations in the U.S. (Beal and others, 1994) were overlain with urban areas from the LULC dataset (U.S. Geological Survey, 1990, 1998). The LULC classifications that were used to define urban areas were "residential," "commercial and services," "industrial," "transportation, communications, and utilities," "industrial and commercial complexes," "mixed urban or built-up land," and "other urban or built-up land."

[24] Selected characteristics include permeability, available water capacity, bulk density, organic matter content, land-surface slope, depth to seasonally high water table, total soil thickness, soil loss tolerance factor, and wind erodibility group.

[25] In 2006, the STATSGO database (U.S. Department of Agriculture, 1994) was revised, resulting in STATSGO2 (U.S. Department of Agriculture, 2011).

[26] Selected characteristics include clay, silt, and sand content; percent of weight of soil material less than 3 inches in size and passing a No. 4, 200, and 10 sieve; percentage calcium carbonate; vertical permeability: permeability of the least permeable soils layer; and cation exchange capacity.

[27] Selected characteristics include hydrologic groups, permeability, available water capacity, bulk density, organic matter content, porosity, soil thickness, field capacity, soil erodibility factor (K factor) of the uppermost horizon, and percent sand, silt, and clay.

[28] Selected characteristics include hydrologic groups, permeability, available water capacity, bulk density, soil restrictive layer, organic matter content, soil erodibility factor (K factor) of the uppermost horizon, and percent sand, silt, and clay.

References Cited

Anderson, J.R., Hardy, E.E., Roach J.T., and Witmer R.E., 1976, A land use and land cover classification system for use with remote sensor data: U.S. Geological Survey Professional Paper 964, 28 p.

Baker, N.T., 2011, Tillage practices in the conterminous United States, 1989–2004—Datasets aggregated by watershed: U.S. Geological Survey Data Series 573, 13 p.

Beal, R.H., Mauldin, J.K., and Jones, S.C., 1994, Subterranean termites—their prevention and control in buildings: U.S. Department of Agriculture, Home and Garden Bulletin no. 64.

Beven, K.J., and Kirkby, M.J., 1979, A physically based, variable contributing area model of basin hydrology: Hydrological Sciences Bulletin, v. 24, p. 43–69.

Clawges, R.M., and Price, C.V., 1999a, Digital data set describing surficial geology in the conterminous United States: U.S. Geological Survey Open-File Report 99–77, accessed April 2003, at http://water.usgs.gov/lookup/getspatial?ofr99-77_geol75m.

Clawges, R.M., and Price, C.V., 1999b, Digital data set describing principal aquifers in the conterminous United States: U.S. Geological Survey Open-File Report 99–77, accessed April 2003, at http://water.usgs.gov/lookup/getspatial?ofr99-77_aquif75m.

Climate Source, Inc., 2006, Spatial climate products: Climate Source, accessed November 2006, at http://www.climatesource.com/products.html.

Conservation Biology Institute, 2006, Protected areas database: Conservation Biology Institute, accessed February 2006, currently at http://consbio.org.

Consortium for International Earth Science Information Network (CIESIN), 1995, Socioeconomic Data and Applications Center (SEDAC) archive of census related products: National Aeronautics and Space Administration (NASA) SEDAC, accessed July 21, 1998, at http://sedac.ciesin.org/plue/cenguide html.

Daly, C., Gibson, W.P., Taylor, G.H., Johnson, G.L., and Pasteris, P., 2002, A knowledge-based approach to the statistical mapping of climate: Climate Research, v. 22, p. 99–113.

Daly, C., and Taylor, G.H., 2002a, Development of new spatial grids of R-factor and 10-yr EI30 for the conterminous United States—Final Report: U.S. Environmental Protection Agency, Las Vegas, Nev., 38 p.

Daly, C., and Taylor, G.H., 2002b, United States mean annual R-factor, 1971–2000: Oregon State University digital raster data, accessed November 2010, at http://prism nacse.org/pub/prism/maps/Precipitation/rfactor/U.S./us_maps.html.

Fenneman, N.M., and Johnson, D.W., 1946, Physiographic divisions of the conterminous United States: U.S. Geological Survey Map, scale 1:7,000,000, digital vector data accessed January 2002, at http://water.usgs.gov/lookup/getspatial?physio.

Foth, H.D., and Schafer J.W., 1980, Soil geography and land use: John Wiley and Sons, Inc., New York, N.Y., 484 p.

Fry, J.A., Coan, M.J., Homer, C.G., Meyer, D.K., and Wickham, J.D., 2009, Completion of the national land cover database (NLCD) 1992–2001 land cover change retrofit product: U.S. Geological Survey Open-File Report 2008–1379, 18 p.

Fry, J.A., Xian, G., Jin, S., Dewitz, J.A., Homer, C.G., Yang, L., Barnes, C.A., Herold, N.D., and Wickham, J.D., 2011, Completion of the 2006 national land cover database for the conterminous United States: Photogrammetric Engineering and Remote Sensing, v. 77, no. 9, p. 858–864.

Gebert, W.A., Graczyk, D. J., and Krug, W. R., 1987, Average annual runoff in the United States, 1951–80: U.S. Geological Survey Hydrologic Investigations Atlas HA–710, scale 1:7,500,000, accessed February 2003, at http://water.usgs.gov/lookup/getspatial?runoff.

GeoLytics, Inc., 2001a, CensusCD 2000/Short form blocks: GeoLytics, East Brunswick, N.J. [digital data on CD-ROM].

GeoLytics, Inc., 2001b, Census 2000 and street 2000: GeoLytics, East Brunswick, N.J. [digital data on 2 CD-ROMs].

Gilliom, R.J., and Thelin, G.P., 1997, Classification and mapping of agricultural land for National Water-Quality Assessment: U.S. Geological Survey Circular 1131, 70 p.

Hamon, W.R., 1961, Estimating potential evapotranspiration: Journal of the Hydraulics Division, Proceedings of the American Society of Civil Engineers, v. 87, p. 107–120.

Hitt, K.J., 1992, 1990 point population coverage for the conterminous United States: U.S. Geological Survey digital vector data, accessed February 15, 2005, at http://water.usgs.gov/lookup/getspatial?uspop90.

Hitt, K.J., 2003, 2000 population density by block group for the conterminous United States: U.S. Geological Survey digital raster data, accessed December 2005, at http://water.usgs.gov/lookup/getspatial?uspopd00x10g.

Hitt, K.J., 2008, Enhanced national land cover data 1992 revised with 1990 and 2000 population data to indicate urban development between 1992 and 2000 (NLCDep0306): U.S. Geological Survey digital raster data, accessed December 2009, at http://water.usgs.gov/lookup/getspatial?nlcdep0306.

Homer, C.G., Dewitz, J., Fry, J.A., Coan, M.J., Hossain, N.D., Larson, C.R., Herold, N., McKerrow, A., VanDriel, J.N., and Wickman, J.D., 2007, Completion of the 2001 national land cover database for the conterminous United States: Photogrammetric Engineering and Remote Sensing, v. 73, no. 4, p. 337–341.

Hunt, C. D., 1979, National atlas of the United States of America—Surficial geology: U.S. Geological Survey, NAC–P–0204–75M–O [map].

Johnson, D.M., and Mueller, R., 2010, The 2009 cropland data layer: Photogrammetric Engineering and Remote Sensing, v. 76, no. 11, pp. 1201–1205.

King, P. B., and Beikman, H.M., 1974a, Explanatory text to accompany the geologic map of the United States: U.S. Geological Survey Professional Paper 901, 40 p.

King, P. B., and Beikman, H.M., 1974b, Geologic map of the United States (exclusive of Alaska and Hawaii) on a scale of 1:2,500,000: U.S. Geological Survey, 3 color plates.

Krug, W.R., Gebert, W.A., and Graczyk, D.J., 1989, Preparation of average annual runoff map of the United States, 1951–80: U.S. Geological Survey Open-File Report 87–535, 414 p.

LaMotte, A.E., 2008a, National land cover database 2001 (NLCD01) tile 1, northwest United States: NLCD01_1: U.S. Geological Survey Data Series 383A, accessed October 2008, at http://water.usgs.gov/lookup/getspatial?nlcd01_1.

LaMotte, A.E., 2008b, National land cover database 2001 (NLCD01) tile 2, northeast United States: NLCD01_2: U.S. Geological Survey Data Series 383B, accessed October 2008, at http://water.usgs.gov/lookup/getspatial?nlcd01_2.

LaMotte, A.E., 2008c, National land cover database 2001 (NLCD01) tile 3, southwest United States: NLCD01_3: U.S. Geological Survey Data Series 383C, accessed October 2008, at http://water.usgs.gov/lookup/getspatial?nlcd01_3.

LaMotte, A.E., 2008d, National land cover database 2001 (NLCD01) tile 4, southeast United States: NLCD01_4: U.S. Geological Survey Data Series 383D, accessed October 2008, at http://water.usgs.gov/lookup/getspatial?nlcd01_4.

Nakagaki, Naomi, 2007a, Grids of agricultural pesticide use in the conterminous United States, 1992: U.S. Geological Survey digital raster data, accessed August 29, 2007, at http://water.usgs.gov/lookup/getspatial?agpest92grd.

Nakagaki, Naomi, 2007b, Grids of agricultural pesticide use in the conterminous United States, 1997: U.S. Geological Survey digital raster data, accessed August 29, 2007, at http://water.usgs.gov/lookup/getspatial?agpest97grd.

Nakagaki, Naomi, Price, C.P., Falcone, J.A., Hitt, K.J., and Ruddy, B.C., 2007, Enhanced national land cover data 1992 (NLCDe 92): U.S. Geological Survey digital raster data, accessed December 1, 2007, at http://water.usgs.gov/lookup/getspatial?nlcde92.

National Atmospheric Deposition Program, 2002, National Atmospheric Deposition Program 2001 annual summary—NADP Data Report 2002–01: Illinois State Water Survey, Champaign, Ill.

National Atmospheric Deposition Program, 2005, National Atmospheric Deposition Program 2004 annual summary—NADP Data Report 2005–01: Illinois State Water Survey, Champaign, Ill.

National Climatic Data Center, 1991, Climate divisions (ed. 1.1): U.S. Geological Survey digital vector data, accessed December 11, 2002, at http://water.usgs.gov/lookup/getspatial?climate_div.

National Climatic Data Center, 2007, Time bias corrected divisional temperature-precipitation-drought index (TD–9640) [digital data files], currently at http://www.ncdc.noaa.gov/oa/climate/climateinventories.html.

National Oceanic and Atmospheric Administration, 1994, NOS80K/ALLUS80K medium-resolution digital vector U.S. shoreline shapefile: National Oceanic and Atmospheric Administration (NOAA), National Ocean Service, Office of Coast Survey and the Strategic Environmental Assessments Division of the Office of Ocean Resources Conservation and Assessment, accessed August 2005, currently at http://woodshole.er.usgs.gov/pubs/of2005-1048/data/basemaps/usa/nos80k/nos80k.htm.

National Oceanic and Atmospheric Administration, 2006, Impervious surface area of the United States: National Oceanic and Atmospheric Administration, National Geophysical Data Center, accessed July 2006 at http://dmsp.ngdc.noaa.gov/html/download_isa2000_2001.html.

Omernik, J.M., 1987, Ecoregions of the conterminous United States. Map (scale 1:7,500,000): Annals of the Association of American Geographers v. 77, no. 1, p.118–125.

Owensby, J.R., and Ezell, D.S., 1992, Climatography of the United States, No. 81—Monthly station normals of temperature, precipitation, and heating and cooling degree days, 1961–90: U.S. Department of Commerce, National Oceanic and Atmospheric Administration, National Climatic Data Center, Ashville, N.C.

Price, C.V., 2003, 1990 Population density by block group for the conterminous United States: U.S. Geological Survey digital raster data, accessed January 24, 2005, at http://water.usgs.gov/lookup/getspatial?uspopd90x10g.

Price, C.V., and Clawges, R.M., 1999a, Digital data sets describing population density in the conterminous United States: U.S. Geological Survey Open-File Report 99–78, accessed May 5, 1999, at http://water.usgs.gov/lookup/getspatial?ofr99-78_popdeng.

Price, C.V., and Clawges, R.M., 1999b, Digital data sets describing metropolitan areas in the conterminous United States: U.S. Geological Survey Open-File Report 99–78, accessed May 4, 1999, at http://water.usgs.gov/lookup/getspatial?ofr99-78_metropop.

Price, C.V., and Clawges, R.M., 1999c, Digital data sets describing toxics release inventory locations with 1995 VOC releases in the conterminous United States: U.S. Geological Survey Open-File Report 99–78, accessed September 27, 2010, at http://water.usgs.gov/lookup/getspatial?ofr99-78_eftri.

Price, C.V., Nakagaki, N., Hitt, K.J., and Clawges, R.M., 2007, Enhanced historical land-use and land-cover data sets of the U.S. Geological Survey: U.S. Geological Survey Data Series 240, accessed August 29, 2007, at http://pubs.usgs.gov/ds/2006/240.

PRISM Climate Group, 2010, Parameter-elevation regressions on independent slopes model (PRISM): PRISM Climate Group, Oregon State University, accessed November 16, 2010, at http://www.prism.oregonstate.edu/.

Radeloff, V. C., Hammer, R. B., Stewart, S. I., Fried, J.S., Holcomb, S. S., and McKeefry, J. F., 2005, The wildland urban interface in the United States: Ecological Applications, v. 15, p. 799–805.

Reed, J.C., and Bush, C.A., 2005, Generalized geologic map of the United States, Puerto Rico, and the U.S. Virgin Islands (ver. 2.0), posted December 2005, accessed September 28, 2010, at http://pubs.usgs.gov/atlas/geologic/.

Ritters, K., Wickham, J., O'Neill, R., Jones, B., and Smith, E., 2000, Global-scale patterns of forest fragmentation: Conservation Ecology, v. 4, no. 2, art. 3.

Rohm, C.M., Omernik, J.M., Woods, A.J., and Stoddard, J.L., 2002, Regional characteristics of nutrient concentrations in streams and their application to nutrient criteria development: Journal of the American Water Resources Association, v. 38, no. 1, p. 1–27.

Ruddy, B.C., Lorenz, D.L., and Mueller, D.K., 2006, County-level estimates of nutrient inputs to the land surface of the conterminous United States, 1982–2001: U.S. Geological Survey Scientific Investigations Report 2006–5012, 17 p.

Schruben, P. G., Arndt, R.E., and Bawiec, W.J., 1998 (release 2), Geology of the conterminous United States at 1:2,500,000 scale—A digital representation of the 1974 P.B. King and H.M. Beikman Map: U.S. Geological Survey Digital Data Series, DDS–11, accessed June 1999, at http://pubs.usgs.gov/dds/dds11.

Schwarz, G.E., and Alexander, R.B., 1995, Soils data for the conterminous United States derived from the NRCS state soil geographic (STATSGO) data base [Original title: State soil geographic (STATSGO) data base for the conterminous United States.]: U.S. Geological Survey Open-File Report 95–449, accessed June 2002, at http://water.usgs.gov/lookup/getspatial?ussoils.

SILVIS Laboratory, 2007, Spatial analysis of conservation and sustainability: SILVIS Laboratory, University of Wisconsin at Madison, Wis., accessed March 2007, at http://www.silvis.forest.wisc.edu/old/maps.php.

Soller, D.R., and Packard, P.H., 1998, Digital representation of a map showing the thickness and character of quaternary sediments in the glaciated United States east of the Rocky Mountains: U.S. Geological Survey Digital Data Series DDS–38, accessed April 16, 2007, at http://pubs.usgs.gov/dds/dds38.

Solley, W.B., Pierce, R.R., and Perlman, H.A., 1998, Estimated use of water in the United States in 1995: U.S. Geological Survey Circular 1200, 71 p., accessed December 10, 2007, at http://water.usgs.gov/watuse/pdf1995/html/.

Steeves, P.A., and Nebert, D.D., 1994, Hydrologic unit maps of the conterminous United States: U.S. Geological Survey digital vector data, accessed January 27, 2005, at http://water.usgs.gov/lookup/getspatial?huc250k.

Thelin, G.P., 2005a, 1992 County pesticide use estimates for 200 compounds (ver. 2.0): U.S. Geological Survey digital file, accessed October 27, 2005, at http://water.usgs.gov/lookup/getspatial?pesticide_use92.

Thelin, G.P., 2005b, 1997 County pesticide use estimates for 220 compounds (ver. 2.0): U.S. Geological Survey digital file, accessed October 27, 2005, at http://water.usgs.gov/lookup/getspatial?pesticide_use97.

Thelin, G.P., 2010, Annual county atrazine use estimates for agriculture (ver. 1.1): U.S. Geological Survey digital tabular data, accessed September 22, 2010, at http://water.usgs.gov/lookup/getspatial?sir2010-5034.

Thorton, P.E., and Running, S.W., 1999, An improved algorithm for estimating incident daily solar radiation from measurements of temperature, humidity, and precipitation: Agricultural and Forest Meteorology, v. 93, p. 211–228.

University of Montana, 2005, Daymet, Daily surface weather and climatological summaries: Numerical Terradynamic Simulation Group, Daymet database, accessed August 2005, currently at http://daymet.ornl.gov.

U.S. Army Corps of Engineers, 2006, National inventory of dams: U.S. Army Corps of Engineers, digital data retrieved July 2006, currently at http://nid.usace.army.mil.

U.S. Bureau of the Census, 1992, Census of population and housing, 1990—Summary Tape file 3A: U.S. Bureau of the Census, Washington, D.C.

U.S. Bureau of the Census, 2001a, 1990 Census block group cartographic boundary files (rev. 4/22/05): U.S. Census Bureau, Geography Division, digital vector data, accessed October 25, 2007, at http://www.census.gov/geo/www/cob/bg1990.html.

U.S. Bureau of the Census, 2001b, 2000 Census block group cartographic boundary files (rev. 4/22/05): U.S. Census Bureau, Geography Division, digital vector data, accessed October 25, 2007, at http://www.census.gov/geo/www/cob/bg2000.html.

U.S. Census Bureau, 2011, TIGER/Line Shapefiles and TIGER/Line Files: U.S. Census Bureau, Geography Division, digital vector data, accessed November 14, 2011, at http://www.census.gov/geo/www/tiger/shp.html.

U.S. Department of Agriculture, 1994, State soil geographic (STATSGO) database—data use information— Miscellaneous publication no. 1492 (rev. ed.): Fort Worth, Texas, Natural Resources Conservation Service [variously paged].

U.S. Department of Agriculture, 1995, 1992 National Resources Inventory: Natural Resources Conservation Service, Washington, D.C., and Statistical Laboratory, Iowa State University, Ames, Iowa [CDROM].

U.S. Department of Agriculture, 1999, 1997 Census of agriculture, Geographic area series 1A, 1B, 1C, Washington, DC., U.S. summary and county level data file: U.S. Department of Agriculture (USDA), National Agricultural Statistics Service (NASS) [producer and distributor, CD–ROM].

U.S. Department of Agriculture, 2000a (Reissued 2001), 1997 National Resources Inventory: Natural Resources Conservation Service, Washington, DC, and Statistical Laboratory, Iowa State University, Ames, Iowa [CD–ROM].

U.S. Department of Agriculture, 2000b, Summary Report: 1997 National Resources Inventory (rev. December 2000): Natural Resources Conservation Service, Washington, DC, and Statistical Laboratory, Iowa State University, Ames, Iowa, 89 p.

U.S. Department of Agriculture, 2000c, Farm resource regions: U.S. Department of Agriculture, Economic Research Service, Agricultural Information Bulletin Number 760, accessed January 11, 2002, currently at http://webarchives.cdlib.org/wayback.public/UERS_ag_1/20111128195215/http:/www.ers.usda.gov/Briefing/ARMS/resourceregions/resourceregions.htm.

U.S. Department of Agriculture, 2004, 2002 Census of Agriculture, Geographic area series, 1A and 1B, Washington, DC, U.S. summary and county level data file: U.S. Department of Agriculture, National Agricultural Statistics Service (NASS) [producer and distributor, CD–ROM].

U.S. Department of Agriculture, 2009, 2007 Census of Agriculture: National Agricultural Statistics Service, accessed January 7, 2009, at http://www.agcensus.usda.gov/.

U.S. Department of Agriculture, 2010a, 2009 Cropland Data Layer: National Agricultural Statistics Service, accessed December 30, 2011, at http:// www.nass.usda.gov/research/Cropland/SARS1a.htm.

U.S. Department of Agriculture, 2010b, Soil survey geographic (SSURGO) database: Soil Survey Staff, Natural Resources Conservation Service, accessed October 18, 2010, at http://soildatamart.nrcs.usda.gov.

U.S. Department of Agriculture, 2011, U.S. general soil map (STATSGO2) database: Soil Survey Staff, Natural Resources Conservation Service, accessed February 24, 2012, at http://soils.usda.gov/survey/geography/statsgo.

U.S. Department of Commerce, 1993, Bureau of the Census, 1:100,000-scale counties of the United States (ed. 1.1): U.S. Geological Survey digital vector data, accessed August 21, 2003, at http://water.usgs.gov/lookup/getspatial?county100.

U.S. Department of Commerce, 1995, 1992 Census of agriculture, geographic area series 1A and 1B, U.S. summary and county level state data: Bureau of the Census, Washington, DC [CD–ROM].

U.S. Environmental Protection Agency, 1998, National strategy for the development of regional nutrient criteria: U.S. Environmental Protection Agency Office of Water, Washington, D.C., 882–R–98–002, 47 p.

U.S. Environmental Protection Agency, 2006, National pollutant discharge elimination system (NPDES): U.S. Environmental Protection Agency, accessed June 2006, at http://cfpub.epa.gov/npdes/.

U.S. Environmental Protection Agency, 2008, National hydrography dataset plus (NHDPlus): U.S. Environmental Protection Agency, U.S. Geological Survey, and Horizon Systems Corporation, accessed October 2008, at http://www.horizon-systems.com/nhdplus.

U.S. Environmental Protection Agency, 2010a, Toxic release inventory (TRI) program fact sheet: U.S. Environmental Protection Agency, accessed October 2010, currently at http://www.epa.gov/tri/triprogram/TRI_factsheeet_Jan_2012.pdf.

U.S. Environmental Protection Agency, 2010b, Level III and IV ecoregions of the continental United States: U.S. Environmental Protection Agency, Western Ecology Division, accessed November 15, 2010, at http://www.epa.gov/wed/pages/ecoregions/level_iii.htm.

U.S. Geological Survey, 1970, The national atlas of the United States of America, Washington, D.C., U.S. Government Printing Office, 417 p.

U.S. Geological Survey, 1990, Land use and land cover digital data from 1:250,000- and 1:100,000-scale maps: U.S. Geological Survey Data User Guide, no. 4, 25 p.

U.S. Geological Survey, 1993, Digital elevation models—data users guide 5: Reston, Va., U.S. Geological Survey, 48 p.

U.S. Geological Survey, 1998, Land use and land cover digital data from 1:250,000- and 1:100,000-scale maps: U.S. Geological Survey, Earth Resources Observation and Science Center, accessed January 1, 2003, at http://edcwww.cr.usgs.gov/products/landcover/lulc.html.

U.S. Geological Survey, 1999a, National land cover data 1992, accessed June 16, 2005, currently at http://www.mrlc.gov/nlcd92_data.php.

U.S. Geological Survey, 1999b, National elevation dataset (NED): U.S. Geological Survey Fact Sheet 148–99, accessed November 16, 2007, at http://erg.usgs.gov/isb/pubs/factsheets/fs14899.pdf.

U.S. Geological Survey, 2001, Elevation Program: U.S. Geological Survey Rocky Mountain Mapping Center, accessed September 28, 2010, at http://rmmcweb.cr.usgs.gov/.

U.S. Geological Survey, 2003, National elevation data set (NED), accessed April 2003, currently at http://nationalmap.gov/elevation.html.

U.S. Geological Survey, 2007a, National land cover database 2001, accessed May 25, 2007, at http://www.mrlc.gov/mrlc2k_nlcd.asp.

U.S. Geological Survey, 2007b, National land cover database (NLCD) 1992/2001 retrofit land cover change product multi-zone download, accessed December 11, 2007, at http://www.mrlc.gov/mrlc2k_multizone.asp.

U.S. Geological Survey, 2011, National land cover database 2006 (NLCD 2006), accessed April 2011, at http://www.mrlc.gov/nlcd06_data.php.

U.S. Geological Survey, 2012, Water Watch --Past streamflow conditions, accessed November 6, 2012, at http://waterwatch.usgs.gov/?id=romap3.

U.S. Geological Survey and U.S. Environmental Protection Agency, 2010, NHDPlus user guide (ver. September 1, 2010): Horizon Systems, accessed September 24, 2010, currently at ftp://ftp.horizon-systems.com/NHDPlusV1/Documentation/NHDPLUSV1_UserGuide.pdf.

Verdin, K.L., and Greenlee, S.K., 1996, Development of continental scale digital elevation models and extraction of hydrographic features. In: Proceedings, Third International Conference/Workshop on Integrating GIS and Environmental Modeling, Santa Fe, New Mexico, January 21–26,1996, National Center for Geographic Information and Analysis, Santa Barbara, Calif. [CD-ROM].

Vista Information Solutions, 1999, StarView real estate user's guide (ver. 2.6.1): Vista Information Solutions, San Diego, Calif., 228 p.

Vogelmann, J.E., Howard, S.M., Yang, L., Larson, C.R., Wylie, B.K., and Van Driel, N., 2001, Completion of the 1990's national land cover data set for the conterminous United States from Landsat Environmental Mapper data and ancillary data sources: Photogrammetric Engineering and Remote Sensing, v. 67, p. 650–662.

Williams, C. N., Menne, M. J., Vose, R. S., and Easterling, D.R., 2006, United States historical climatology network daily temperature, precipitation, and snow data, ORNL/CDIAC–118, NDP–070: Carbon Dioxide Information Analysis Center, Oak Ridge National Laboratory, Oak Ridge, Tennessee, accessed November 16, 2010, at http://cdiac.ornl.gov/epubs/ndp/ushcn/ndp070.html.

Wolock, D.W., 1997, STATSGO soil characteristics for the conterminous United States: U.S. Geological Survey Open-File Report 97–656, accessed March 14, 2001, at http://water.usgs.gov/lookup/getspatial?muid.

Wolock, D.M., 2003a, Base-flow index grid for the United States: U.S. Geological Survey Open-File Report 03–263, accessed September 9, 2005, at http://water.usgs.gov/lookup/getspatial?bfi48grd.

Wolock, D.M., 2003b, Saturation overland flow estimated by TOPMODEL for the conterminous United States: U.S. Geological Survey Open-File Report 03–264, accessed April 16, 2007, at http://water.usgs.gov/lookup/getspatial?satof48.

Wolock, D.M., 2003c, Estimated mean annual natural groundwater recharge in the conterminous United States: U.S. Geological Survey Open-File Report 03–311, accessed April 16, 2007, at http://water.usgs.gov/lookup/getspatial?rech48grd.

Wolock, D.M., 2003d, Infiltration-excess overland flow estimated by TOPMODEL for the conterminous United States: U.S. Geological Survey Open-File Report 03–310, accessed April 16, 2007, at http://water.usgs.gov/lookup/getspatial?ieof48.

Wolock, D.M., 2003e, Hydrologic landscape regions of the United States: U.S. Geological Survey Open-File Report 03–145, accessed December 3, 2004, at http://water.usgs.gov/lookup/getspatial?hlrus.

Wolock, D.M., and McCabe, G.J., 1995, Comparison of single and multiple flow direction algorithms for computing topographic parameters in TOPMODEL: Water Resources Research, v. 31, p.1315–1324.

www.ingramcontent.com/pod-product-compliance
Lightning Source LLC
Chambersburg PA
CBHW082031190526

45166CB00017B/2939